Competition and Investment in Air Transport

Ruwantissa Abeyratne

Competition and Investment in Air Transport

Legal and Economic Issues

Springer

Ruwantissa Abeyratne
Global Aviation Consultancies Inc.
Montreal
Québec, Canada

ISBN 978-3-319-24371-9 ISBN 978-3-319-24372-6 (eBook)
DOI 10.1007/978-3-319-24372-6

Library of Congress Control Number: 2015958724

Springer Cham Heidelberg New York Dordrecht London

Printed on acid-free paper

Springer International Publishing AG Switzerland is part of Springer Science+Business Media (www.springer.com)

Preface

Here are some basic facts.

Half the world is living in cities. It is not probable that globalization will stop and, with exponential development, this city population will only grow globally. An OXFAM report published in January 2015 states that 1 % of the world's richest people own 42 % of the total global wealth while the other 99 % own the balance 58 %. Of these, the world's richest 20 % own almost all of the 58 % leaving just 5.5 % to almost 80 % of the world. In just 2 years, the top 1 % will have more wealth than the other 99 % if this trend continues.[1] An ageing population, many with disposable incomes, is another irreversible trend. Against this backdrop, increasing urbanisation; expanding middle class; and rise in migration, tourism and international students are current and future trends.

Amidst these revealing figures, a PWC Global Airline CEO Survey conducted in 2014 states that airline CEOs expect three trends to dominate over the coming years: technological advances; shifts in the global economy, and demographic changes that would transform their businesses.[2] Development and international cooperation is a buzz word in many developed and developing countries. Air travel will double in 2035 as against today's figure. In January 2015, ongoing projects for airport construction amounted to the value of US $543 billion globally.

These facts and figures incontrovertibly spell out the future of air transport and the inevitable fact that liberalization of air transport is a compelling need to meet demand. However, protectionism of market access is looming its head, taking us back to the frustrating 1970s and 1980s. Technology is changing rapidly, affecting the way air transport is being conducted around the world.

Globalization and deregulation are no longer intrinsically linked to each other.

[1] *Wealth: Having It All and Wanting More*, OXFAM: January 2014 at 1.

[2] *Strategic Sights Set on Transformation and Innovation*, PWC Global Airline CEO Survey 2014, www.pwc.com/transport at 1.

Although the prevailing cross border flow of people will increase, the quantum of cross border investment of foreign direct investment would probably remain at the current rate of around 10 %, thus attracting continued protectionism in air transport.[3] The World Bank, in its January 2015 Report,[4] expects overall global growth to rise moderately to 3.0 % in 2015, and average about 3.3 % through 2017. The Report posits that a growth rate of 2.2 % will be seen in high income countries in 2015–2017, which would be an increase of 1.8 % as against 2014, on the back of gradually recovering labour markets, ebbing fiscal consolidation, and still low financing costs.

Growth is projected to gradually accelerate in developing countries, rising from 4.4 % in 2014 to 4.8 % in 2015 and 5.4 % by 2017.

Competition in air transport is integrally connected to and impacted by various drivers such as environment and climate change, foreign trade and investment, economic growth of nations, liberalization and the enhancement of national security, and even the commercialization of outer space activities, which is in turn affected by globalization—a phenomenon which opened trade barriers through politics and diplomacy. Integration of the world trading machine into a globally connected process occurred with the twin forces of globalization and the information revolution. Air transport is hedged between these two factors, attenuating the advantages offered by both.

In early July of 2015, the 3-year long study on the expansion of London's Heathrow Airport was released with the recommendation that a third runway be built at Heathrow to accommodate traffic growth and promote London as a growing global hub. The main drivers of this recommendation were that London is lagging behind other European hubs such as Schipol and Frankfurt, which were offering significantly more flights out of Amsterdam and Frankfurt to China and other emerging markets in Asia, and that Dubai's airport capacity has exceeded the total airport capacity of all of Britain, thus weakening Britain's connectivity with the key points of economic growth in the world that offer opportunities for trade. The simple fact is that, unless British air transport facilities and infrastructure are brought to par with the rest of its competitors, Britain will be out of the competitive field. This brings to bear the incontrovertible need for capacity growth with which the intrinsic link between competition in air transport and connectivity could be strengthened.

Also at the time this book was being written, the five major airlines of Europe (Air France KLM, EasyJet, International Airlines Group, Lufthansa Group and Ryanair) met with a view to seeking a new European policy on air transport. They suggested *inter alia* that there should be a more airline-friendly liberalization policy in Europe that could seek reductions in airport charges at major airports through tighter regulation, which in turn would save consumers €1.5 billion in lower fares. Airport Council International—the voice of airports worldwide—countered that the suggestion was ludicrous and extraordinary because it hinged on one key assumption—that

---

[3] Pankaj Ghemawat, *World 3.0: Global Prosperity and How to Achieve it*, Harvard Business Review Press: 2011 at 29.

[4] *Global Economic Prospects: Having Fiscal Space and Using it*, January 2015, at 21.

any such reduction in airport charges would be passed on 100 % by airlines to passengers. ACI further contended that independent research showed that airline interest does not necessarily equal consumer interest and that to assume airlines would pass mandated reductions in airport charges on to consumers was ludicrous to say the least. ACI further complained that most airlines do not even effectively refund airport charges and other ticket taxes to passengers who do not take their flight.

Air transport has certainly come full circle where competition is rife. During its incipient days, the days of PANAM, TWA, KLM and QANTAS in the 1950s and early 1960s, flying was a luxury, where passengers dressed up for the trip as if they were attending a wedding and sat in lounge chairs throughout their journey enjoying gourmet food. Then came the transition to cost cutting and coach class where airlines were packed like sardines with mostly the budget complying tourist and start-up businessman. First class was a rare commodity reserved for the few rich and corporate high ups. Now, in the early twenty-first century, we seem to be back where we started. Of course, coach class is still the major component, but the opulent ostentation with which first class and business class passengers are treated and the facilities they enjoy both in the air and on the ground simply boggles the mind.

For the air traveller who can afford it (mostly in instances where his/her company pays for the trip), the buzz words for airlines are "passenger experience". There are sleeping rooms and a cigar lounge in Lufthansa's first class lounge. Airlines such as Lufthansa and Qatar Airways drive their first class passengers right up to the plane in a limousine. Foot massages, six senses spas and even acupuncture and facial treatments are standard features at most first class airline lounges. Some services have personal butler services, three roomed cabins with 32 inch televisions on the wall and an in-house chef who can turn out anything you fancy. It is reported that in Cathay Pacific's Wing lounge—reserved for first class passengers—the arm chairs are designed by an Italian architect and the walls of the lounge are eucalyptus. There are private cabanas, showers, beds and even a soaking tub. At Virgin Atlantic's premium check-in at London Heathrow, first class passengers can drive right up to the terminal, sit comfortably in a plush reclining chair while their documentation is processed, be fast tracked through security and whisked off to the Clubhouse Lounge, which is equipped with a *mixologist* who will mix your drink to your liking; a Jacuzzi, roof deck and salon. Of course, one cannot forget the black truffle martini. This goes way beyond the standard VIP check-in, chauffeur service and lounge that is standard in business class.

Virgin is reported to have a Revival Lounge that trusses you up when you arrive, which offers you a shower, a clothes pressing service while you shower and a chauffeur driven car waiting for you to take you home. Abu Dhabi airport has even a shoe shine service and a barber, all at your disposal to give you a fresh, wet shave!

Tyler Dickman, CEO at Lounge Buddy which is an app that provides information on global availability of airport lounges has commented: "your biggest risk is forgetting that you are in an airport and missing your flight", as reported recently in the Montreal Gazette.

So why is all this happening? Thomas Piketty, in his bestselling book *Capital in the Twenty First Century* says that inequality began to rise sharply in the 1970s and

1980s, which brings us to the present where half of the population of the world own nothing; the poorest 50 % own less than 10 % of national wealth and generally less than 5 %. The richest 10 % of the world command 62 % of the total wealth while the poorest 50 % own only 4 %. The basis for this trend is the equation r > g (where *r* represents the rate of return on one's capital and *g* represents the growth of the economy). Piketty, who is a professor at the Paris School of Economics says: "indeed, in the United States, as in France and Europe, today, as in the past, income from capital always becomes more important as one climbs the rungs of the income hierarchy…the upper centile (the richest 1 %) consists of several social groups, some with very high incomes from capital and others with very high incomes from labour". Taking the United States as an example, Piketty concludes that "it is perfectly possible to imagine that the top centile's share of total wages could reach 15–20 % or 25–30 % or even higher".

With regard to r > g, when the rate of return on capital exceeds the growth rate it invariably results in increased inequality in terms of distribution of wealth. If the growth rate (g) is 1 % and the rate of returns is 10, the investor has to invest only a tenth of his/her annual capital to ensure the growth of his/her annual capital income over the average income. This economic fact has led to rising and continuing inequality in the world where, in colloquial and clichétic terms, the rich get richer and the poor get poorer.

It is said, although not accepted by economists as a rational yardstick, that 50 % of the world is poor, 40 % belongs to the middle class and 10 % is wealthy, of which 1 % is super rich. As was said earlier, the richest 10 % of the world command 62 % of the total wealth while the poorest 50 % own only 4 %, which gives one a perspective of why airlines have got back to bending backwards in competing with their rivals for excellence in first and business class service. Etihad Airways describes its first class service in its Airbus A380 called "The First Apartment" as follows: "*The First Apartment is upholstered with the finest Poltrona Frau leather and features a wide armchair, an ottoman and even includes space to walk around in. The ottoman opens out into an 81-inch long bed.*

A companion can join you for a meal or meeting in your spacious First Apartment. When travelling with a partner the divider between Apartments can be lowered to join rooms.

The First Apartments come equipped with an adjustable 24-inch flat screen TV, which can be viewed from the armchair or while relaxing in bed, together with a vanity cabinet with ample lighting and space for your personal effects, a chilled drinks cabinet and a wardrobe."

With all this fanfare, air transport is a contentious industry mainly due to the fact that there is intense competition among air carriers and the countries they represent as national carriers. The conflict between the United States carriers and the carriers of the United Arab Emirates is a case in point which is discussed in some detail in this book.

Montreal, QC, Canada Ruwantissa Abeyratne
September 2015

Contents

1 The Future of Air Transport Law . . . 1
1.1 Introduction . . . 1
1.2 Where Is the World Headed? . . . 3
1.3 Challenges Facing Air Transport Law . . . 6
1.3.1 Nationalism and Sovereignty . . . 6
1.3.2 Market Access . . . 8
1.3.3 The Role of ICAO in Air Transport . . . 9
1.3.4 Commercial Space Transport . . . 9
1.3.5 Rights of the Passenger . . . 11
References . . . 13

2 Competition in Aerospace Travel . . . 15
2.1 Introduction . . . 15
2.2 International Cooperation in the Peaceful Uses of Outer Space . . . 20
2.3 Definition and the Delimitation of Outer Space . . . 26
2.3.1 The Antarctic Treaty . . . 29
2.3.2 The Maritime Analogy . . . 31
References . . . 35

3 Competition and the Environment . . . 37
3.1 Introduction . . . 37
3.2 The Work of ICAO . . . 42
3.2.1 The Environment Action Group (EAG) . . . 42
3.2.2 The CAEP Process . . . 44
3.3 The Strawman Approach . . . 45
3.3.1 Application of the *Strawman* Approach to MBMs . . . 47
3.3.2 *Strawman* Considerations for ICAO . . . 48
References . . . 51

4 Competition and Safety . . . 53
4.1 Introduction . . . 53
4.2 Impact of Competition on Safety . . . 53

4.3 Safety Information 55
4.3.1 Considerations of the Conference 56
4.3.2 Recommendations 57
4.4 Discussions at the Second High Level Safety Conference 58
4.4.1 Protection of Safety Information 58
4.4.2 Legal Issues Involved 61
4.4.3 GermanWings 63
4.4.4 Dissemination and Sharing of Safety Information 66
4.4.5 Legal Issues 68
4.4.6 Consumer Protection 69
References 72

5 Competition and Innovation in Air Transport 73
5.1 Introduction 73
5.1.1 Competition 73
5.1.2 Principles of Competition 78
5.1.3 Legal Issues 87
5.1.4 Past, Present and the Future 98
5.1.5 The UAE Carriers 99
5.1.6 The Mathematical Theories 102
5.2 Market Economics and Air Transport 103
5.3 Competition and Innovation 107
References 110

6 Market Access and Subsidies in Air Transport: The US-UAE Debate and WTO 113
6.1 Introduction 113
6.1.1 The Debate 113
6.1.2 The United States and Europe 118
6.2 Market Access and WTO 121
6.2.1 ICAO Perspectives 121
6.2.2 Some Misconceptions 124
6.3 Subsidies and WTO 126
References 130

7 Achieving Competitive Advantage Through Connectivity and Innovation: An Application in Airline Hubbing 131
7.1 Introduction 131
7.2 The Need for Connectivity in Air Transport 134
7.3 Hubbing 136
7.4 Innovation 139
References 142

8 Competition and Open Skies 145
8.1 Introduction 145
8.2 A Case Study of ASEAN and Indonesia 150

8.3 Integration into the Global Economy 153
8.3.1 The Global Economy 153
8.3.2 ASAM and Air Transport Economics in Indonesia 158
References 163

9 The NAFTA Free Trade Model 165
9.1 Introduction 165
9.2 Economic Environment of NAFTA Parties 167
9.2.1 Canada 167
9.2.2 The United States 170
9.2.3 Mexico 172
9.3 Historical and Economic Aspects of NAFTA: Background and Rationale 176
9.4 Legal Principles Governing NAFTA 181
9.4.1 The Legal Status of NAFTA at International Law 181
9.4.2 Rights and Obligations of Parties to NAFTA 184
References 188

10 Competition Law in Air Transport 191
10.1 Introduction 191
10.2 Competition Laws in Air Transport 193
10.2.1 State Aid 196
10.2.2 Dominant Position 199
10.2.3 Competition Laws in the United States 201
10.2.4 Some Conclusions 202
References 206

11 Air Transport and the Law of Investment 207
11.1 Introduction 207
11.2 Competition and Investment in Air Transport Infrastructure . . . 209
11.3 Attribution 211
References 214

12 Conclusion 215

Appendix A: Relevant Provisions of Treaty on the Functioning of the European Union on State Aid 223
Aids Granted by States 223
Article 107 223
Article 108 224
Article 109 225

Appendix B: Relevant Competition Provisions of the Clayton Act 227
§ 2 Clayton Act, 15 U.S.C. §§ 13(2) 227
§ 3 Clayton Act, 15 U.S.C. § 14 229

Appendix C: The Sherman Antitrust Act (1890) 231

Index 233

Chapter 1
The Future of Air Transport Law

1.1 Introduction

Half the world is living in cities. It is not probable that globalization will stop and, with exponential development, this city population will only grow globally. An ageing population, many with disposable incomes, is another irreversible trend. Against this backdrop, increasing urbanisation; expanding middle class; and rise in migration, tourism and international students are current and future trends. Development and international cooperation is a buzz word in many developed and developing countries.

Air travel will double in 2035 as against today's figure. It is forecast that, between 2009 and 2028 there will be a demand for 24,951 passenger and freighter aircraft worth US $3.1 trillion, and that, by 2028 there will be 32,000 aircraft in service compared with 15,750 in 2009. In January 2015, ongoing projects for airport construction amounted to the value of US $543 billion globally. In April/ May 2015, The International air Transport Association (IATA) recorded that worldwide airline share prices were up 12 % in April as against the figure in the same month in 2014 and global air travel rose by a robust 5.9 % for the same period. Air freight volumes were up by 3.3 % as well.

These facts and figures incontrovertibly spell out the future of air transport and the inevitable fact that liberalization of air transport is a compelling need to meet demand. However, protectionism of market access is looming its head, taking us back to the frustrating 1970s and 1980s.

Additionally, technology and development resonate the fact that in the near future, commercial space travel will take off, posing a challenge to tenets of air transport law. Regrettably, in all these areas, initiatives in air transport law have been insouciant at best.

At the outset a seminal point of nomenclature has to be clarified. The law pertaining to air transport has been quoted, and often misquoted, with the use of various terms, the first being "air law". *Milde* asserts that the term is "controversial

R. Abeyratne, *Competition and Investment in Air Transport*,
DOI 10.1007/978-3-319-24372-6_1

and imprecise", saying that it is possibly influenced by the French term "*droit et aerostats*"[1] and that it is misleading in that it wrongfully implies a separate branch of law. He asserts that the term "aeronautical law" would be more to the point although the sustained use of the term "air law" should be respected.[2] According to *Milde*, there is no autonomy that could be ascribed to the term "air law" which he claims is composed of various principles that applies to social relations (physical persons, corporate bodies and sovereign States) regulated by law. Although one cannot question this premise, one could certainly question whether the term "aeronautical law" as suggested by *Milde* as the alternative, is consistent with his argument about social relations, since aeronautical law is essentially the law relating to "aeronautics" which is defined as "a science that deals with airplanes and flying" or the science dealing with the operations of aircraft.[3]

Dempsey, on the other hand is seemingly comfortable with the term "air law" when he says that that international air law or aviation law is composed of public and private categories[4] and that the *Magna Carta* of air law is the Chicago Convention of 1944.[5] Be that as it may, the author prefers to use the term "air transport law" for purposes of this book.

Another disturbing trend that has been consistent throughout the past 70 years is that the meaning and purpose of law as it applies to air transport has been upended, in that the law has inflexibly dictated changing economic and social circumstances instead of the other way around, where the law, as a management and social tool, should adapt to changing circumstances and be changed accordingly. At the forefront of this inequity are two key influences or drivers—international law and politics—and their treatment of sovereignty of States and air traffic rights which concepts have not overcome the antiquated notions of market share and protectionism that belie modern exigencies of market economics. Generally under legal theory, each State is sovereign and equal and the term *sovereignty* may be used as a synonym for independence. However, in modern parlance, with the rapid growth in telecommunications and global competition and rivalries, no State can be entirely sovereign to the exclusion of others.

Inextricably connected to this phenomenon is the awareness of the international community that, within the progression of air law are two main issues concerning the evolving role of air law. The first is that that the distinction between air law and space law is continuing to blur. The second is that principles of air law are getting increasingly involved in activities related to military warfare. With regard to the threat envisioned in the use of military warfare on civil aviation activity, the same players, i.e. international law and politics, play the same role. Although airspace is

---

[1] Milde (2012), p. 1.

[2] *Ibid.*

[3] http://www.merriam-webster.com/dictionary/aeronautics.

[4] Dempsey (2015), p. 215.

[5] Convention on International Civil Aviation, signed at Chicago on 7 December 1944. See ICAO Doc 7300/8:2006.

common and States have sovereignty over the airspace above their territories, this does not enable them to use such air space arbitrarily. There are strict principles with regard to aerial military activity and prohibition of the use of military warfare on civilian populations and properties. These must be strictly adhered to in the basis of political consensus.

Air law and space law are closely inter-related in some areas and both these disciplines have to be viewed in the twenty-first century within the changing face of international law and politics. Both air law and space law are disciplines that are grounded on principles of public international law, which is increasingly becoming different from what it was a few decades ago. We no longer think of this area of the law as a set of fixed rules, even if such rules have always been a snapshot of the law as it stands at a given moment.

With changing technology, old political dogma and economic theory are no longer viable and today's challenges demand that we look at the world in a new way. With changing political philosophy, where governments are increasingly asserting their sovereignty, a whole new compromise in air transport law becomes necessary. Although one speaks of globalization, in practice, when one looks at cross border integration of markets, the world is only semi-globalized with barriers that effectively preclude air transport's full potential. Air transport, more than any other industry, demonstrates that a borderless world is still a theoretical concept where State interest still takes prominence over the consumer.

New and emerging threats to civil aviation will continue to be a cause for concern to the aviation community. Grave threats such as those posed by the carriage of dangerous pathogens on board, the use of cyber technology calculated to interfere with air navigation systems, and the misuse of man portable air defence systems will remain real and will have to be addressed with vigour and regularity.

Another area that requires attention is the reactive and ineffective manner the legal regime applicable to air transport was put to work over 2014, where several air disasters took the world by surprise. The regulators scrambled to set things right, and it was as though they had not imagined that such disasters were possible.[6]

It is submitted that the enduring weakness of air transport law is the disconnect between where the world is headed and the role to be played by air transport as a product that should be a more efficient engine for growth.

1.2 Where Is the World Headed?

For the next few years at least, the global order portends a disturbing uncertainty. Economic power is shifting across the globe to emerging markets in the Far East. Technology is changing rapidly, affecting the way air transport is being conducted

[6] See Abeyratne (2014a). Also see Abeyratne (2014b), pp. 329–342; and Abeyratne (2014c), pp. 544–558.

around the world. Globalization and deregulation are no longer intrinsically linked to each other. Although the prevailing cross border flow of people will increase, the quantum of cross border investment of foreign direct investment would probably remain at the current rate of around 10 %, thus attracting continued protectionism in air transport.[7] The World Bank, in its January 2015 Report,[8] expects overall, global growth to rise moderately, to 3.0 % in 2015, and average about 3.3 % through 2017. The Report posits that a growth rate of 2.2 % will be seen in high income countries in 2015–2017, which would be an increase of 1.8 % as against 2014, on the back of gradually recovering labour markets, ebbing fiscal consolidation, and still low financing costs. Growth is projected to gradually accelerate in developing countries, rising from 4.4 % in 2014 to 4.8 % in 2015 and 5.4 % by 2017.

A significant gap in the shape of things to come and existing air transport law is the disconnect between where the world is heading, both politically and technologically, and the laws needed to steer air transport in line with shifting trends while offering a safer, more secure and more efficient product. The power shift to the east, where consumer spending in China is $2.2 trillion in 2015 and Middle Eastern countries such as Saudi Arabia, Qatar UAE, Kuwait and Bahrain have $1 trillion in investment, are harbingers of the direction the world economy is taking. These figures have to be read in conjunction with some basic facts on the direction air transport is headed. Covering the years 1980–2013 a study[9] was conducted by the OECD,[10] which reflects that the airline sector is continuing to grow exponentially. Another compelling fact The OECD Report brings to bear is that by 2026, air transport will contribute $1 trillion to world's GDP.[11] The International Civil Aviation Organisation (ICAO)[12] posits that passenger trips increased from 4.028 billion in 1980 to 19.125 billion in 2012 and that international scheduled passenger traffic grew by 5.2 % in 2013 in comparison to 2012 and is expected to reach over

[7] Ghemawat (2011), p. 29.

[8] *Global Economic Prospects*: *Having Fiscal Space and Using it*, January 2015 at 21.

[9] AIRLINE COMPETITION—Background Paper by the Secretariat, Directorate for Financial and Enterprise Affairs Competition Committee, 18–19 June 2014, DAF/COMP(2014)14.

[10] Organisation for Economic Co-operation and Development (OECD), established in 1961, promotes policies that are calculated to improve the economic and social well-being of people around the world. The OECD provides a forum in which governments can work together to share experiences and seek solutions to common problems. The Organization works with governments to understand what drives economic, social and environmental change.

[11] AIRLINE COMPETITION, *Supra* note 9 (this chapter), *Background Note* at 3. The Report goes on to say that worldwide, aviation and related tourism generate over 56 million jobs, of which 8.36 million are directly linked to the aviation sector. Around 35 % of international tourists travel by air.

[12] The International Civil Aviation Organization is the United Nations specialized agency dealing with international civil aviation. ICAO was established by the Convention on International Civil Aviation (Chicago Convention), signed at Chicago on 7 December 1944. See ICAO Doc 7300/8: 2006. The main objectives of ICAO are to develop the principles and techniques of international air navigation and to foster the planning and development of air transport. ICAO has 191 Contracting States.

6.4 billion passenger by 2030.[13] According to ICAO's forecast, there will be an average annual growth rate of 4.5 % by 2030 in passenger traffic (of both scheduled and unscheduled services).[14]

The key drivers of air transport, according to an *Airbus Industrie* forecast, of economic growth will be: increasing urbanisation; expanding middle class; and rise in migration, tourism and international students.[15] This forecast predicts that emerging countries-regions (Asia and the Pacific, Africa, Middle East and South America) will overtake the developed countries-regions in terms of economic growth with a 10 % increase in growth in passenger travel.

There are three areas that would be crucial in the years to come if we are to avoid self-induced stagnation. They are: competition for growth; international intervention to secure the welfare of people; and investment in a balanced education and healthcare for the people. As for competition for growth, this is not a new measure of economic proactivity.

The World Economic Forum reports in its Global Agenda[16] that geostrategic competition is a compelling sign of future global trends and that recent developments have led to tectonic shifts in state interaction, bringing to centre stage geopolitics and realpolitik causing wide ranging effects on the world economy. Air transport, which has remained a political tool in view of the legal recognition of sovereignty in air space, would undoubtedly be affected by this trend.

In view of the above, when one applies the trajectory of the global economy and its direction in the coming years to the market economics of the air transport industry, it becomes eminently clear that the economic forces that are shaping the global economy will affect the progress of aviation. The World Energy Council (WEC) has reported that fuel demand in the transport sector in the next 40 years will come mainly from developing countries such as China and India, where demand will grow by 200–300 %. In contrast, the WEC is of the view that the transport fuel demand for the developed countries will drop by up to 20 %, mainly due to increased efficiencies. The demand of the developing countries is expected to surpass that of the developed countries by the year 2025.

The report also forecasts that oil may still fuel more than 80 % of the global transport sector for the next 40 years due to strong demand growth from the heavy duty sector, shipping and air traffic. WEC projects that by 2050 global fuel demand in all transport modes could increase by 30–82 % compared to 2010 levels. This portends the inevitability that fossil fuels, the reserves of which are still being

[13] ICAO Press Release, 16 December 2013. The OECD Report also notes that the number of travellers has increased because, among many other things, prices have decreased significantly in response to increasing competition in the air transport market. For example, in 1974 the cheapest round-trip New York-Los Angeles flight (in inflation-adjusted dollars) that regulators would allow: $1442. Today one can fly that same route for $268.

[14] Global Air Transport Outlook to 2030, *Circ.333*, *AT/190*:2012 at 59.

[15] Airbus Industrie, Global Market Forecast: Flying on Demand 2014–2033 at 16.

[16] World Economic Forum: outlook on the Global Agenda 2015, http://www.weforum.org/reports/outlook-global-agenda-2015.

discovered, will retain its heavy influence over the coming years and therefore global efforts would have to be concentrated on market based measures as well as the development of alternative fuel technology.

Against this backdrop, and in view of the air transport forecasts discussed above, an IATA forecast which predicts that air travel will double over the next 20 years[17] becomes extremely relevant, inevitably bringing to bear a dichotomy—that in the absence of a more liberalized air transport regime than what prevails currently, this exponentially increased air travel market could be stultified.

1.3 Challenges Facing Air Transport Law

1.3.1 Nationalism and Sovereignty

With regard to the direction in which the world is headed and the impact it has on air transport, arguably, the most significant future challenge to air transport law would be rising nationalism and sovereignty in air space, where the latter has been misunderstood by many States to confer to them absolute immunity against their domestic decisions and actions. This misconception has been exacerbated by Article 1 of the Chicago Convention which recognizes that States have complete and exclusive sovereignty over the airspace above their territory. In light of this lack of clarity in air transport law, many States still believe that air transport services should be subservient to their parochial national interests of protectionism.

The Economist states that it would become necessary in 2015 to recognize that nationalism is back. The trend in politicians' agenda would be to claim that they are standing up for their own countries and this would cut across Europe, Asia and the Americas. The political ambition behind this strategy would be popularity that would enable politicians to grow in power. *The Economist* is of the view that there will be an increase in international tensions and an "unpromising background for efforts at multilateral co-operation, whether on climate, trade, taxation or development[18]".

Nationalism, when merged with sovereignty of air space forms a dangerous combination that presents a misconception that ascribes primacy to protectionism. This would be detrimental to the "fair and equal opportunity" for carriers to operate air services and compete that is provided for in the Preamble to the Chicago Convention. Moreover, it would form an ominous cocktail with both Article 1 and Article 6 of the Convention, the latter of which requires an airline operating

[17] *The Shape of Air Travel Markets Over the Next 20 Years*, https://www.iata.org/publications/economics/Pages/index.aspx.

[18] Nationalism is back: Bad news for international co-operation, *The Economist: The World in 2015* at 92. See the web version of Nov 20th, 2014 at http://www.economist.com/news/21631966-bad-news-international-co-operation-nationalism-back.

scheduled air services to obtain permission from a grantor state to fly in and out of its territory, and would undermine the spirit of globalization and its very purpose of economic efficiency.

Steinberger in 2000 observed that numerous and varied legal obligations of States prescribed by international legal instruments would essentially preclude them from exercising the puritanical concept of sovereignty, thus constraining their actions. It was *Steinberger's* view that such international responsibility would endorse a State's sovereignty rather than diminish it.[19] *Amitai Etzioni*, in a compelling article speaks of the new idea of sovereignty which was endorsed by UN Secretary General Kofi Annan, that there is a radical turnaround from the concept of sovereignty (absolute and exclusive rights of States within their borders which no other State could question or interfere with) of the Peace of Westphalia of 1648, to where sovereignty is not absolute but conditional, where a State could only maintain its sovereignty on condition that it met with its national and international obligations.[20]

Etzioni's compelling argument can be traced to UN Secretary General *Kofi Annan*'s statement before the United Nations General Assembly in 1999 where he said that State sovereignty was being redefined in its most basic sense by the forces of globalization and international cooperation. Annan said that the State was "widely understood to be the servant of its people, and not *vice versa*".[21] In similar vein, *Starke* is inclined to stretch the principle of sovereignty to accommodate external involvement by a State in the affairs of another in special circumstances:

> ..."Sovereignty" has a much more restricted meaning today than in the eighteenth and nineteenth centuries when, with the emergence of powerful highly nationalized States, few limits on State autonomy were acknowledged. At the present time there is hardly a State which, in the interests of the international community, has not accepted restrictions on its liberty of action.[22]

If, as the foregoing discussion reflects, air travel will double by 2035 and several thousands of new aircraft are placed in the market, ICAO will have to attempt a definition of sovereignty or in the least offer an interpretation that would be consistent with modern exigencies of market economics in air transport. This could be accommodated in a Repertory Guide to the Chicago Convention that is long overdue.[23]

---

[19] Steinberger (2000), p. 501.

[20] Etzioni (2005–2006), p. 35.

[21] Secretary-General Presents his Annual Report to General Assembly, 20 September 1999, Press Release, SG/SM/7136, GA/9596. See http://www.un.org/press/en/1999/19990920.sgsm7136.html.

[22] Starke (1977), p. 106.

[23] In the 1970s ICAO made a half-hearted attempt at developing a Repertory Guide to the Chicago Convention and has done nothing since toward explaining the legal and regulatory interpretation of the various provisions of the Convention. A commentary of the Convention was developed by the author in 2013, which analyses and discusses the key provisions of the Convention. See generally Abeyratne (2013a).

1.3.2 Market Access

The ambiguity brought about by the concepts of sovereignty in air space as well as the restriction imposed by Article 6 of the Chicago Convention has given rise to some confusion between open skies on the one hand and protectionism on the other. This has impacted on the financing of airlines, which the Chicago Convention has nothing to do with. The recent spat between the carriers of the United States and the carriers of the UAE and Qatar are a case in point. The term "subsidy" is not defined precisely in economic terms although the *Oxford English Dictionary* defines the word as "a sum of money granted from public funds to help an industry or business keep the price of a commodity or service low". In broad terms therefore, a subsidy can be considered indirect protectionism. Under the WTO *Agreement on Subsidies and Countervailing Measures* (SCM Agreement),[24] a subsidy is recognized as a financial contribution by a government which confers a benefit.[25] A financial contribution is either money or anything else of value provided to a manufacturer or exporter (which could be construed as an international air service originating in a country) at a cost less than would have been charged in a commercial transaction. This could include indirect support from a government.[26]

It is quite obvious that the air transport industry stands at the crossroads of two major influences—globalization and the information revolution—which have revolutionized the trading world by driving competition. The fact that the UAE carriers (as well as Qatar Airways and Turkish Airlines) have the geographic advantage of being in the centre of the long air routes of the world should not be taken as an argument supporting competitive distortion or disadvantage to other carriers which would in turn unduly curb the operations of the four carriers. The Air Transport Association (IATA)[27] has recognized that US $4.2 trillion is needed over the next two decades to properly offer connectivity to a travelling public who are creating an exponential demand for air transport.[28]

IATA's Director General Tony Tyler has said that what is needed is smart regulation from governments around the world in order to maximize the economic benefits of connectivity—jobs and growth.

The World Trade Organization has clear rules on subsidies as a trade barrier but market access in air transport is not included within the purview of the WTO. This creates a challenge for both ICAO and air transport law, to introduce either guidelines or rules on this subject. As the discussion below suggests, ICAO,

[24] https://www.wto.org/english/tratop_e/scm_e/subs_e.htm.

[25] Article 1.1.

[26] Cunningham (1999), p. 6.

[27] The International Air Transport Association is a trade association of the world's airlines. These 250 airlines, primarily major carriers, carry approximately 84 % of total available air traffic.

[28] Schaal (2015). See http://skift.com/2013/07/01/the-airline-business-is-a-terrible-one-says-leading-airline-industry-group/.

being itself confused on its role in air transport, may not be of much help unless it straightens out its own act in the field of air transport.

1.3.3 The Role of ICAO in Air Transport

The role of ICAO in the field of air transport is at best ambivalent and ICAO has been recognized for what it is rather than what it has done. Although one of ICAO's strategic objectives is *Economic Development of Air Transport* which is to foster the development of a sound and economically viable civil aviation system, ICAO goes on to say that the Strategic Objective reflects the need for ICAO's leadership in harmonizing the air transport framework focused on economic policies and supporting activities. It does not take a legal expert to recognize that this entire statement is fundamentally flawed, convoluted and meaningless. Firstly it says that ICAO fosters the sound and economical a civil aviation system. "Civil aviation system" encompasses the entirety of civil aviation (not merely air transport) including aircraft engine and frame manufacturers and component manufacturers. It then goes on to say that there is a need for ICAO's "leadership" in harmonizing (not "fostering" this time) of the air transport framework (with no mention of the civil aviation system). It is no surprise that the ICAO Council, which doubtless was prompted by the Secretariat to adopt these words, has maundered through confusion worse confounded. This unsatisfactory state of affairs has been documented in depth elsewhere.[29]

Over the next few years, ICAO would, as a matter of necessity, have to address the issue of the role of air transport law as it should apply globally, even if such were to end up in a lesser compelling code of conduct. In the least, the prevailing ambivalence would have to be clarified with a way forward for ICAO to assert its much vaunted and claimed "leadership" role in harmonizing the air transport framework which is now relegated to a regime of serf serving fragmentation. When everything is said and done, ICAO's role in this area amounts to much said and nothing done.

1.3.4 Commercial Space Transport

It would not be incorrect to say that air transport stands at the threshold of crossing over to the realm of commercial space transport, particularly since the latter is designed to use aerospace craft which begin their journey from ground through the atmosphere. Thus, air transport law will come under careful scrutiny in the future. Also, the question as to whether the Annexes to the Chicago Convention would have to be adapted would become relevant. The recent development of the SABRE

[29] Abeyratne (2013b), pp. 9–29. Also see Abeyratne (2013c), pp. 297–332.

engine which could reduce engine temperature from 1500 to −150 °C at re-entry into the atmosphere by the aerospace plane would reduce a journey between London and Sydney to just 4 h. It is claimed by the SABRE engine manufacturer that with this new development, a journey from anywhere to anywhere else in the world could not take more than 4 h at the most. This would indeed impact on competition between carriers who would be using aerospace transport equipment and conventional long haul carriers.

There seems to be no doubt that some ICAO involvement will be necessary, in view of the inevitable overlap between the air transport and space transport segments of a journey, particularly in the field of air traffic management, and analogous principles may have to be attenuated from the Annexes to the Chicago Convention on subjects such as licensing, documentation to be carried on board, and rules of outer space travel.

There are a number of provisions of the Chicago Convention that could be adapted to apply to commercial space transport. For example, Article 37 which calls for State cooperation in ensuring harmonization of their regulations with those of the Annexes to the Convention can, in a space law context, provide that each State undertakes to collaborate in securing the highest practicable degree of uniformity in regulations, standards, procedures, and organization in relation to spacecraft personnel, space paths and auxiliary services in all matters in which such uniformity will facilitate and improve commercial space transportation. The exception to this requirement as contained in Article 38 can also be adapted to afford some flexibility to any State which finds it impracticable to comply in all respects with any such international standard or procedure, or to bring its own regulations or practices into full accord with any international standard or procedure after amendment of the latter, or which deems it necessary to adopt regulations or practices differing in any particular respect from those established by an international standard. In such an instance a State could give immediate notification to the organization empowered with the regulatory direction to be provided to States of the differences between its own practice and that established by the international standard. In the case of amendments to international standards, any State which does not make the appropriate amendments to its own regulations or practices could be required to give notice to the monitoring body within 60 days of the adoption of the amendment to the international standard, or indicate the action which it proposes to take. Another provision that can be meaningfully adapted is Article 34 which could provide that in respect of every spacecraft engaged in commercial space transport a journey log book in which shall be entered particulars of the spacecraft, its crew and of each journey, in such form as may be prescribed from time to time pursuant to the legal instrument covering commercial space transport. In any such case, the monitoring body could be empowered to l make immediate notification to all other States of the difference which exists between one or more features of an international standard and the corresponding national practice of that State.[30]

[30] For an extensive discussion on this subject, see generally Abeyratne (2014d).

1.3.5 Rights of the Passenger

Of all the key areas that affect the future of air transport law, the rights of the passenger on a global scale are the most contentious, purely because there simply are no globally recognized legal rights of the passenger. Curiously, the legal constitution of air transport—the Chicago Convention—is all about States and air carriers with no mention of the consumer of the air transport product except in a few isolated instances that are linked to the functions of a State and an air carrier.[31] The European Union leads the way with its passenger rights policy. The EU passenger rights policy is quite equitable and protects the passenger well. Under this policy, a passenger may not be charged a higher price for a ticket because of his nationality or where he buys the ticket from. Secondly, he has certain rights in case things go wrong. This concerns delays, cancellations and overbooking that prevent him from boarding if he is departing from any airport situated in the EU, or arriving in the EU with an EU carrier or one from Iceland, Norway or Switzerland. There are obligations cast upon the carrier in instances of denied boarding or cancelation of flight or overbooking. In such instances the passenger is entitled to either: transport to the passenger's final destination using comparable alternative means, or having his ticket refunded and, where relevant, being returned free of charge to his initial departure point. If the flight is delayed by 5 h or more, the passenger is also entitled to a refund. The airline must inform the passenger about his rights and the reason for being denied boarding, or any cancellations or long delays (over 2 h, although this may be up to 4 h for flights in excess of 3500 km). With regard to compensation for denied boarding, or if the flight is cancelled or arrives more than 3 h late on arrival at the final destination stated on the passenger's ticket, he may be entitled to compensation of €250–600, depending on the distance of the flight, if it is within the EU for an amount of €250 involving 1500 km or less, and €400 for a trip over 1500 km. This amount differs if the carriage was to be between an EU airport and a non-EU airport which is €250 for a trip involving 1500 km or less; €400 for a trip of 1500–3500 km; and €600 for over 3500 km.[32]

However, mostly in the rest of the world it has been left to individual States with their policies on consumer protection to look after air passengers who are often not aware of their rights (if there are any) in a particular jurisdiction.[33] The preeminent fundamental human right of a passenger is the right not to be wronged. His rights, when taken collectively, act as the fundamental postulate of any constitutional democracy. Not only must a right remedy a wrong that has been committed or obviate a wrong that ought to be prevented, but a right must be contrived according to circumstance and formally recognized.

[31] For a discussion on the rights of the passenger see Abeyratne (2015), pp. 159–275.

[32] http://europa.eu/youreurope/citizens/travel/passenger-rights/air/index_en.htm.

[33] http://www.cbc.ca/news/politics/airline-passengers-in-dark-about-rights-advocate-says-1.2489336.

ICAO is just beginning to focus some attention to what it calls consumer protection and was scheduled at the time of writing, to conduct a seminar later in 2015 on the subject. It would be counter intuitive and counter-productive if what comes out of this event are merely some recommendations or conclusions. A serious discussion on globally accepted rules and a code of conduct for air carriers and States are warranted.

It is unfortunate that, in the twenty-first century and after 70 years of regulated air transport, there is no such thing as a globally accepted set of air transport laws. What exists is a sporadic discussion of isolated issues in a global forum which is confused about its own role in air transport. It is time that this fragmented discussion crystallized into global acceptance by States of at least the key aspects of air transport that have been the subject of this discussion. For this to happen, two key elements are necessary. The first would be the conduct by the Council of ICAO under Article 55 c) of the Chicago Convention[34] of a study of the future of air transport and the laws which could address problems that are looming. An example of an ominous threat to air transport was recently reported where a cyber security consultant had hacked into computer systems aboard airliners up to 20 times and managed to control an aircraft engine during a flight. The consultant, apparently with the good intention of alerting authorities against the threat of hackers manipulating aircraft's flight pattern, had also hacked into in-flight entertainment systems aboard aircraft. He claimed to have done so 15–20 times from 2011 to 2014. He had also revealed an alarming fact—that he had once hacked into the systems and then overwrote code, enabling him to issue a "CLB," or climb, command.[35]

The second step would be to extend the study of the Council to an inquiry into the feasibility of submitting compelling rules of a globally acceptable nature or codes of conduct (COCs) with regard to issues that may have consequences for air transport from a global perspective. The incorporation of such rules or COCs into local laws would be the final accomplishment. As per Lord Atkin:

> International law has no validity except in so far as its principles are accepted and adopted by our own domestic law. . .The courts acknowledge the existence of a body of rules which nations accept among themselves. On any judicial issue they seek to ascertain what the relevant rule is, and having found it they will treat it as incorporated into the domestic law, so far as it is not inconsistent with rules enacted by statutes or finally declared by their tribunals.[36]

---

[34] Article 55 c) contains one of the permissive functions of the Council of ICAO which provides that the Council may conduct research into all aspects of air transport and air navigation which are of international importance and communicate such to member States of ICAO while facilitating the exchange of information between the States.

[35] http://venturebeat.com/2015/05/17/fbi-says-this-hacker-took-over-a-plane-through-its-in-flight-entertainment-system/.

[36] *Chung Chi Cheung* v. *R*, [1939] AC 160; 9 AD, p. 264. See also *Commercial and Estates Co. of Egypt* v. *Board of Trade* [1925] 1 KB 271, 295; 2 AD at 423.

The key to meeting the challenges posed by air transport in the future is development and international cooperation. For these two accomplishments to attain fruition, a robust and dynamic set of air transport laws would be essential.

References

Abeyratne R (2013a) The convention on international civil aviation – a commentary. Springer, Heidelberg

Abeyratne R (2013b) Reinventing ICAO's role in air transport – a compelling need? Issues Aviat Law Policy 13(1):9–29

Abeyratne R (2013c) The air transport conference of ICAO – a critical analysis. Air Space Law XXXVII(4–5):297–332

Abeyratne R (2014a) Flight MH 17 and state responsibility for ensuring safety and security of air transport. J Air Transport Secur. doi:10.1007/s12198-014-0148-0, published on line 2 October 2014

Abeyratne R (2014b) Flight MH 17: the legal and regulatory fallout. Air Space Law 39 (6):329–342

Abeyratne R (2014c) Flight MH 370 and global flight tracking – the ICAO reaction. Zeitschrift fur Luft- und Weltraumrecht (German J Air Space Law) 63:544–558

Abeyratne R (2014d) Regulation of commercial space transport – the astrocizing of ICAO. Springer, Heidelberg

Abeyratne R (2015) Aviation and international cooperation: human and public policy issues. Springer, Heidelberg, pp 159–275

Cunningham RO (1999) Subsidies to large civil aircraft production: new WTO subsidy rules and dispute settlement mechanism alter dynamics of U.S.–E.U. dispute. Air Space Law 6

Dempsey PS (2015) The future of air law in the 21st century. Zeitschrift fur Luft- und Weltraumrecht (German J Air Space Law) 64(2):215

Etzioni A (2005–2006) From right to responsibility, the definition of sovereignty is changing. Insight 35

Ghemawat P (2011) World 3.0: global prosperity and how to achieve it. Harvard Business Review Press, p 29

Milde M (2012) In: Benko M (ed) International air law and ICAO, 2nd edn. Eleven International Publishing, The Hague, p 1

Schaal D (2015) The airline business is a lousy one, airline industry study shows. SKIFT: Megatrends Defining Travel in 2015

Starke JG (1977) An introduction to international law, 7th edn. Butterworth, London, p 106

Steinberger H (2000) Sovereignty. In: Bernhardt R (ed) Encyclopaedia of public international law, vol IV. Elsevier, Amsterdam, p 501

Chapter 2
Competition in Aerospace Travel

2.1 Introduction

As stated in the previous chapter, air transport law would have an inextricable influence on commercial space transport, and it would be a matter of time before competition in aerospace travel becomes an inevitable corollary to competition law in air transport. The two systems—commercial air transport and commercial space transport would be linked through space ports and facilities offered as well as sovereignty issues in aerospace transport which, at the time of writing this book, were ambivalent and vague at best. Therefore, it becomes necessary to devote a chapter to some aspects of aerospace transport, particularly in terms of some legal nuances that would impact of a merged aerospace regime.

The preeminent consideration for competition both for air transport and space transport would be State control of air space which a commercial aerospace-craft traverses. The protectionism that still prevails in competition in this area would continue to play a role in terms of air traffic rights of an aerospace plane, particularly because the limitations of air space and outer space have not been defined nor recognized as a critical element in anticipation of the inevitable expansion of the commercial space travel market. It will be a matter of time when the aerospace plane enters the commercial aerospace transport market not just for space tourism but for commercial transport. In such an instance, in a few years, competition in this travel market will rise, and at the center of contention would be the delimitation of air space and outer space, which still remains unresolved. How would the first and second freedoms of the air be interpreted in the context of orbital travel of the aerospace plane. In this context, efforts and non-efforts of the international community in determining the limits of the heavens would become relevant if a clear regulatory approach to competition in aerospace travel were to be envisioned.

In early 2013 The Committee on the Peaceful Uses of Outer Space of the United Nations (COPUOS) prepared a paper on States responses to one of three questions asked of them, on whether their governments had given consideration to the

R. Abeyratne, *Competition and Investment in Air Transport*,
DOI 10.1007/978-3-319-24372-6_2

possibility of defining a lower limit of outer space and/or upper limit of airspace, recognizing at the same time the possibility of enacting special international legislation relating to a mission carried out by an object in both airspace and outer space. The reason for this is obvious; there is no clear agreement, since the launch of Sputnik in 1957, 58 years on, as to what constitutes airspace and at what point in altitude outer space starts. What the international community is contending with are two approaches to the issues: the functionalist approach and the spatialist approach. In the face of exponential strides made in commercial space travel, these two ambivalent paths are woefully inadequate by themselves to charter a legal and regulatory way forward.

As the discussion in this chapter reflects, The United Nations, in addition to five multilateral treaties on the exploration of outer space, has been consistently active in laying out principles and guidelines on space exploration through various Resolutions of the General Assembly. All these efforts would be more meaningful if they were presented against the backdrop of a clear understanding of what constitutes outer space and at what point it begins attitudinally.

At the 3rd Manfred Lachs International Conference of the Institute of Air and Space Law, McGill University, on *New Space Commercialization and Law* held on 16–17 March 2015 in Montreal, Dr. Ram Jakhu of the Institute highlighted the distance the international community has travelled from 1957—when Sputnik was launched—to the modern concept of *New Space* which he identified as a movement composed of several themes: disruptive technology, entrepreneurship; privately held companies; individual access to and settlement of space and exploitation of space reserves. Two days later, at the ICAO/UNOOSA Aerospace Symposium on the theme: *Emerging Space Activities and Civil Aviation – Challenges and Opportunities* held in Montreal from 18 to 20 March 2015 Dr. Simonetta de Pippo, Director, United Nations Office for Outer Space Affairs said that the inter-relationship and dialogue between major space faring nations and emerging space nations relating to increased international cooperation and capacity-building efforts for the benefit of developing countries has continued to ensure a link for success, and that the space agenda was continuously evolving and becoming more complex, one of the reasons for which was the expanding commercial space sector. Dr. de Pippo concluded by saying that space tools were fundamental to meeting the challenges to humanity and sustainable development and the overarching space security environment in its broader sense caters for global space governance.

The statements of both these eminent scholars bring to bear the incontrovertible prospect of the important role of space law in an increasingly evolving area of transportation that would define the years to come. They also point to the significance of governance in space seemingly within the parameters of delimitation and demilitarization. Delimitation, or the boundary between air space and outer space, has been both a contentious issue and a matter of sardonic conjecture which prompted scholars to call it a comedy of errors.[1] There is an ongoing debate as to

[1] Cheng (1980), p. 323.

whether the delimitation should be based on functionalism i.e. the activity indulged in, or spatialism, i.e. an altitudinal cap where outer space starts and air space ends. Years ago the prevalent view was that eventually the issue would transform itself into favouring the former, as was seen both in a scholarly perspective[2] and in the world of the professionals as evidenced in the 1959 Report of the UN ad hoc COPUOS.[3]

One of the key discussions to international cooperation in outer space activities is the prevention of weaponization of outer space, on which numerous agreements were entered into in the 1960s and 1970s, the most prominent of which are the Partial Test Ban Treaty[4] and the Outer Space Treaty.[5] The story goes as far back as 20 November 1959, where the United Nations General Assembly, by Resolution 1378 (XIV) called upon Governments to make every effort to achieve a constructive solution of the problem of general and complete disarmament, and expressed the hope that measures leading towards the goal of general and complete disarmament under effective international control would be worked out in detail and agreed

---

[2] *Ibid.*

[3] In 1959, The General Assembly of the United Nations established the Committee on the Peaceful Uses of Outer Space as a permanent body through its Resolution 1472 (XIV)—*International co-operation in the peaceful uses of outer space*. In 1961, the General Assembly, considering that the United Nations should provide a focal point for international cooperation in the peaceful exploration and use of outer space, requested the Committee, in cooperation with the Secretary-General and making full use of the functions and resources of the Secretariat; to maintain close contact with governmental and non-governmental organizations concerned with outer space matters; to provide for the exchange of such information relating to outer space activities as Governments may supply on a voluntary basis, supplementing, but not duplicating, existing technical and scientific exchanges; and to assist in the study of measures for the promotion of international cooperation in outer space activities. The 1959 Report of the United Nations ad hoc Committee on the Peaceful Uses of Outer Spaces appeared to favour functionalism. See Cheng (1980), p. 324.

[4] Treaty Banning Nuclear Weapon Tests in the Atmosphere, in Outer Space and Under Water, 5 August 1963 Signed at Moscow by the Original Parties, the Union of Soviet Socialist Republics, the United Kingdom of Great Britain and Northern Ireland and the United States of America. The Treaty was opened for signature at London, Moscow and Washington on 8 August 1963 and entered into force on 10 October 1963.

[5] Treaty on Principles Governing the Activities of States in the Exploration and Use of Outer Space, including the Moon and Other Celestial Bodies, United Nations, *Treaty Series*, Vol. 610, No. 8843. The Treaty was opened for signature by the three depository Governments (the Russian Federation, the United Kingdom and the United States of America) in January 1967, and it entered into force in October 1967. The Outer Space Treaty provides the basic framework on international space law that works to establish that: the exploration and use of outer space shall be carried out for the benefit and in the interests of all countries and shall be the province of all mankind; outer space shall be free for exploration and use by all States; outer space is not subject to national appropriation by claim of sovereignty, by means of use or occupation, or by any other means; States shall not place nuclear weapons or other weapons of mass destruction in orbit or on celestial bodies or station them in outer space in any other manner; the Moon and other celestial bodies shall be used exclusively for peaceful purposes; astronauts shall be regarded as the envoys of mankind; States shall be responsible for national space activities whether carried out by governmental or non-governmental entities; States shall be liable for damage caused by their space objects; and States shall avoid harmful contamination of space and celestial bodies.

upon in the shortest possible time, Arguably one of the most significant pronouncements on the issue was in United Nations General Assembly Resolution 52/37 which reaffirmed that the legal regime applicable to outer space by itself does not guarantee the prevention of an arms race in outer space.[6]

Fundamental to international cooperation in the exploration of outer space is demilitarization. This is based on the basic premise that it is the will of States that the exploration and use of outer space, including the Moon and other celestial bodies, shall be for peaceful purposes and shall be carried out for the benefit and in the interest of all countries, irrespective of their degree of economic or scientific development and Article 2 of the United Nations Charter that States must refrain from using threat and force against other States. In Resolution 52/37 of December 1997,[7] The United Nations General Assembly reaffirmed the importance and urgency of preventing an arms race in outer space, and the readiness of all States to contribute to that common objective, in conformity with the provisions of the Outer Space Treaty and called upon States, in particular those with major space capabilities, to contribute actively to the objective of the peaceful use of outer space and of the prevention of an arms race in outer space and to refrain from actions contrary to that objective and to the relevant existing treaties in the interest of maintaining international peace and security and promoting international cooperation.

In December 2001 The General Assembly adopted Resolution 56/23[8] wherein the General Assembly emphasized the paramount importance of strict compliance with existing arms limitation and disarmament agreements relevant to outer space, including bilateral agreements, and with the existing legal regime concerning the use of outer space, and noted that that wide participation in the legal regime applicable to outer space could contribute to enhancing its effectiveness. There were earlier Resolutions which conveyed the same message,[9] as well as later Resolutions on the same subject of prevention of an arms race in outer space.[10]

Against the backdrop of both delimitation and demilitarization comes the emergent issue of suborbital flights[11] and space tourism, where commercial enterprises

---

[6] A/RES/52/37, 67th Plenary Meeting, 23 December 1997, Prevention of an Arms Race in Outer Space, Resolving Clause 2.

[7] A/RES/52/37, 23 December 1997, Prevention of an arms race in outer space.

[8] A/RES/56/23, 21 December 2001, Prevention of an arms race in outer space.

[9] See A/RES/55/32, 3 January 2001 as well as A/RES/32, 3 January 2001; CA/RES/54/53, 31 December 1999; A/RES/53/76, 4 January 1999.

[10] See A/RES/57/57, 30 December 2002; A/RES/58/36, 8 January 2004; A/RES/59/65, 17 December 2004; A/RES/60/54, 6 January 2006; A/RES/61/58, 3 January 2007; and A/RES/62/20, 10 January 2008.

[11] A sub-orbital flight is a flight up to a very high altitude which does not involve sending the vehicle into orbit. Sub-orbital trajectory, which a sub orbital flight would follow, is defined in the legislation of the United States as "The intentional flight path of a launch vehicle, re-entry vehicle, or any portion thereof, whose vacuum instantaneous impact point does not leave the surface of the Earth." In 2004, SpaceShipOne was the first private vehicle to complete two sub-orbital flights

are already developing space vehicles that can not only take tourists on sub orbital flights—such as Virgin Galactic's SpaceShipTwo and Sierra Nevada Corporation's Dream Chaser which is a reusable crewed suborbital and lifting body space plane, but also the aerospace plane of Mitsubishi Heavy Industries which is being developed to minimize flying time, for example, from Los Angeles to Tokyo in 2 h and from anywhere in the world to anywhere else within just 4 h. One of the issues that sub orbital flights raises is whether, at the height the flights are conducted, the vehicle is deemed to be in air space or outer space. Therefore, sub orbital flights inevitably call for a determination as to what might be air space, as against outer space.

Another issue that brings to bear the significance in the current context of the subject of delimitation is the portentous relevance of mining for asteroids. It is but a matter of time that private commercial entities take on this practice. As a corollary the two critical issues that emerge are: who is responsible for and accountable when private companies or individuals conduct outer space activities and who could claim ownership of samples when they are brought to earth. If trillions of dollars worth of platinum and cobalt (not to mention other precious resources) are mined from asteroids nudged to near earth orbit, could there be private ownership of this property. In response to the first question on responsibility, The Outer Space Treaty provides that States Parties bear international responsibility for national activities in outer space, including the moon and other celestial bodies, whether such activities are carried out by governmental agencies or non-governmental entities, and for assuring that national activities are carried out in conformity with the provisions set forth in the treaty. The Treaty further states that the activities of non-governmental entities in outer space, including the moon and other celestial bodies would in essence inevitably require authorization and continuing supervision by the appropriate State Party.

As for the second question of ownership, the Outer Space Treaty prescribes that outer space is a resource that belongs to all mankind. Therefore it follows that property which resides in outer space belongs to all. However, economic theory suggests that property rights and claims thereto emerge when it is in someone's self interest to claim property, and that claims to such rights are prompted by desires of States, governments or individuals purely based on cost benefit possibilities. In this context, one has to wait and see what will develop in this expensive but valuable exercise worth trillions of dollars.

It is in consideration of the above issues that this article attenuates the historical development of international cooperation in the peaceful uses of outer space that is calculated to help humankind not only in space transportation but also in various

within 2 weeks carrying weight equivalent to three human adults up to about 62.5 miles (100 km) to win the Ansari X Prize. It was carried for 1 h by an aeroplane up to nearly 50,000 feet (9.5 miles) from where it was released into a glide and then propelled vertically for 80 s by a rocket motor to an altitude of more than 62 miles at apogee, reaching a speed over Mach 3. Then falling back to return to earth, it re-entered the atmosphere and glided during 15–20 min before landing back on the runway of departure.

aspects of life on earth. It also traces the various aspects of outer space activities identified by United Nations Resolutions and evaluate where we started and where we are now. Finally, there is a discussion on analogical examples that could prove to be of interest in the way forward in straightening out issues of delimitation.

2.2 International Cooperation in the Peaceful Uses of Outer Space

Originally, The *ad hoc* Committee on the Peaceful Uses of Outer Space (UNCOPUOS or COPUOS as has already been mentioned) was established in 1958 by Resolution 1348 (XIII)[12] to report on the area of international co-operation and programmes in the peaceful uses of outer space which could appropriately be undertaken under United Nations auspices to the benefit of States irrespective of the state of their economic or scientific development, taking into account *inter alia*: continuation on a permanent basis of the outer space research then being carried on within the framework of the International Geophysical Year; organization of the mutual exchange and dissemination of information on outer space research; co-ordination of national research programmes for the study of outer space, and the rendering of all possible assistance and help towards their realization; the future organizational arrangements to facilitate international co-operation in this field within the framework of the United Nations; and; the nature of legal problems which may arise in the carrying out of programmes to explore outer space.

UNCOPUOS proper was established by General Assembly Resolution 1472 to review, as appropriate, the area of international cooperation, and to study practical and feasible means for giving effect to programmes in the peaceful uses of outer space which could appropriately be undertaken under United Nations auspices in the area of assistance for the continuation on a permanent basis of the research on outer space carried on within the framework of the International Geophysical Year; organization of the mutual exchange and dissemination of information on outer space research; and encouragement of national research programmes for the study of outer space and rendering of all possible assistance and help toward their realization. An important additional task assigned to UNCOPUOS was to study the nature of legal problems which may arise from the exploration of outer space.[13]

In 1961, as per Resolution 1721 B (XVI) The General Assembly resolved that the United Nations should provide for international cooperation in the peaceful exploration of outer space. In examining in November 1961 the Report of the

---

[12] A/RES 1348 (XIII) Question of the peaceful use of outer space, *792nd plenary meeting*, 13 December 1958.

[13] A/RES/1472 (XIV), International Cooperation in the peaceful uses of outer space, 856th plenary meeting, 12 December 1959, Resolving Clause 1.

COPUOS the United Nations General Assembly in Resolution 2260 (XXII) expressed that the Assembly was convinced that the widest possible cooperation in the exploration and use of outer space can be an important contributing factor to friendly relations among States. The Resolution also called upon States launching objects into orbit or beyond to furnish information promptly to UNCOPUOS, for the registration of launchings.[14]

The first instance of recognition that there must be more awareness of human environment and that space technology could be a vital tool in this understanding, the General Assembly, in 1969 adopted Resolution 2600 (XXIV) which invited member States with experience in the field of remote earth resources surveying to make such experience available to other member States which do not have such experience and encourage them to become familiar in that respect. The Resolution also invited member States to join in exploring the various aspects involved in the analysis of data obtained through earth resources surveying techniques, their dissemination and application, so as to maximise the benefits to be obtained therefrom, taking into account the particular interests and needs of developing countries.[15] This is a seminal step taken by the United Nations in establishing the need for developed countries to offer their expertise in space technology for the benefit of less privileged or less developed countries, which is the bedrock of international cooperation in outerspace affairs.

In 1970 the General Assembly went further, where Resolution 2733 (XXV) recognized that the use of satellite-borne television for educational and training purposes, particularly in developing countries, can in many instances contribute towards national programmes of integration and community development and economic, social and cultural developments in such areas as formal and adult education, agriculture, health and family planning.[16] The Resolution also expressed the importance of potential benefits derived from direct broadcast satellite services, especially in developing countries and invited the International Telecommunications Union (ITU) to continue to take necessary steps to promote the use of satellite broadcasting services by member States.[17]

In 1971 the United Nations General Assembly, through its Resolution 2776 (XXVI) welcomed the efforts of numerous member States to share with other interested member States the practical benefits that may be derived from programmes in space technology.[18] The Assembly also welcomed efforts of the Food and Agriculture Organization (FAO)—in promoting international cooperation

---

[14] A/RES/1721 (XVI) B, International Cooperation in the peaceful uses of outer space, 1085th plenary meeting, 20 December 1961, Resolving Clause 1.

[15] A Res/2600 XXIV, International cooperation in the Peaceful uses of outer space, 186th Plenary, 16 December 1969, Resolving Clauses 1 and 2.

[16] A RES/27333 (XXV) International cooperation in the peaceful uses of outer space, 1932nd plenary meeting, 16 December 1970.

[17] *Id.* Resolving Clauses 2 and 6.

[18] A/RES/2776 (XXVI) International Cooperation in the peaceful uses of outer space, 1998th plenary meeting, 29 November 1971, Resolving clause 9.

in education and training in the peaceful uses of outer space—the World Meteorological Organization (WMO)—in mobilizing technical resources in order to discover ways and means in mitigating the harmful effects and destructive potential of tropical storms—and the United Nations Educational, Scientific and Cultural Organization (UNESCO) and ITU—in satellite broadcasting for the purpose of contributing to the advancement of education and training. Finally, The International Atomic Energy Agency (IAEA) was requested to provide, together with other relevant international Organizations, to provide COPUOS progress reports on their work relating to the peaceful uses of outer space.[19] This trend established by the General Assembly was reiterated in Assembly Resolution A/3182 (XXVII) which was adopted in December 1973.[20]

The gradual pace at which the United Nations General Assembly moved in drawing in the various key players in their role of cooperation in the peaceful uses of outer space was fluid and purposeful. This trend continued relentlessly, weaving a tapestry of mutual cooperation and assistance. A new dimension was added in 1978 where the General Assembly requested the continuation of work by the Scientific and Technical Sub Committee of UNCOPUOS on: issues relating to remote sensing of the earth by satellites; consideration of the United Nations programme on space applications and the coordination of space activities within the United Nations system; questions of convening a United Nations conference on outer space; and questions relating to space transportation systems.[21]

Over the next years many resolutions followed, progressing step by step in the peaceful exploration of outer space.[22] A landmark resolution of the UN General Assembly was Resolution 41/64 adopted in December 1986 which called for fellowships to be awarded for the study of questions relating to remote sensing of the earth by satellites; use of nuclear power sources in outer space; questions relating to space transportation systems and their implications for future activities in space; and the examination of the physical nature and technical attributes of geostationary orbit.[23]

[19] *Id.*, Resolving Clauses 15, 16, and 17.

[20] A/RES/3182, International cooperation in the peaceful uses of outer Space, 2205th plenary meeting, 18 December 1973.

[21] A/RES/33/16, International cooperation in the peaceful uses of outer space, 51st plenary meeting, 10 November 1978, Resolving Clause 6.

[22] A/RES//RES/4/66, 89th plenary meeting, 5 December 1979; UNGA/RES/36/35, 63rd Plenary meeting, 18 November 1981; A/RES/38/80, 98th Plenary meeting, 15 December 1983; A/RES/39/96, 100th Plenary meeting, 14 December 1989; A/RES/41/162, 118th Plenary meeting, 16 December 1985 (it is to be noted that this resolution called, inter alia, for the use of techniques by all member States of techniques resulting in medical studies conducted in space), A/RES/42/68 (which drew attention to micro gravity experiments in space and their application), 89th Plenary meeting, 2 December 1987; UNGA/RES/43/56, 71st Plenary meeting, 6 December 1988.

[23] A/RES/41/64, International cooperation in the peaceful uses of outer space, 95th Plenary meeting, 3 December 1986.

In December 2006 The United Nations General Assembly adopted Resolution 61/75 on transparency and confidence-building measures in outer space activities inviting all Member States to submit to the Secretary-General of the United Nations before its 62nd Session concrete proposals on international outer space transparency and confidence-building measures in the interest of maintaining international peace and security and promoting international cooperation and the prevention of an arms race in outer space.[24] This was followed by Resolution 62/217 of February 2008 which *inter alia* exhorted the Legal Committee of COPUOS to address, while considering the special needs of developing countries, matters relating to: the definition and delimitation of outer space; the character and utilization of the geostationary orbit, including consideration of ways and means to ensure the rational and equitable use of the geostationary orbit without prejudice to the role of the International Telecommunication Union; review and possible revision of the Principles Relevant to the Use of Nuclear Power Sources in Outer Space; examination and review of the developments concerning the draft protocol on matters specific to space assets to the Convention on International Interests in Mobile Equipment; capacity-building in space law; consider the general exchange of information on national legislation relevant to the peaceful exploration and use of outer space in accordance with the work plan adopted by the Committee.[25]

In Resolution 63/90 adopted in December 2008 the General Assembly further requested the Legal Committee of UNCOPUOS to address matters relating to the definition and delimitation of outer space; and the character and utilization of the geostationary orbit, including consideration of ways and means to ensure the rational and equitable use of the geostationary orbit without prejudice to the role of the International Telecommunication Union. The Committee was also requested to consider for discussion the review and possible revision of the Principles Relevant to the Use of Nuclear Power Sources in Outer Space; examination and review of the developments concerning the draft protocol on matters specific to space assets to the Convention on International Interests in Mobile Equipment; capacity-building in space law; and consideration of general exchange of information on national mechanisms relating to space debris mitigation measures.[26]

In 2010, one of the aspects focused by the General Assembly was collision of space objects. In Resolution 64/86 the Assembly considered it essential that Member States pay more attention to the problem of collisions of space objects, including those with nuclear power sources, with space debris, and other aspects of space debris, calls for the continuation of national research on this question, for the development of improved technology for the monitoring of space debris and for

[24] A/RES/61/75, 67th Plenary meeting, 6 December 2006. See also A/RES/68/50, Transparency and confidence-building measures in outer space activities, 68th Session, 10 December 2013.

[25] A/RES/62/217, International cooperation in the peaceful uses of outer space, 62nd Session, 1 February 2008, Resolving Clause 6.

[26] A/RES/63/90, International cooperation in the peaceful uses of outer space, 63rd Session, 18 December 2008 Resolving clause 4.

the compilation and dissemination of data on space debris, also considers that, to the extent possible, information thereon should be provided to the Scientific and Technical Subcommittee, and agrees that international cooperation is needed to expand appropriate and affordable strategies to minimize the impact of space debris on future space missions.[27] The Resolution also urged all States, in particular those with major space capabilities, to contribute actively to the goal of preventing an arms race in outer space as an essential condition for the promotion of international cooperation in the exploration and use of outer space for peaceful purposes.[28]

In 2012, by its Resolution 66/71, the General Assembly asserted its deep belief that the use of space science and technology and their applications in areas such as tele-health, tele-education, disaster management, environmental protection and other Earth observation applications contribute to achieving the objectives of the global conferences of the United Nations that address various aspects of economic, social and cultural development, particularly poverty eradication.[29] The Assembly reaffirmed the importance of international cooperation in developing the rule of law, including the relevant norms of space law and their important role in international cooperation for the exploration and use of outer space for peaceful purposes, and of the widest possible adherence to international treaties that promote the peaceful uses of outer space in order to meet emerging new challenges, especially for developing countries. The Resolution reiterated focus of Resolution 64/86 on the problem of collision of space objects and again urged all States, in particular those with major space capabilities, to contribute actively to the goal of preventing an arms race in outer space as an essential condition for the promotion of international cooperation in the exploration and use of outer space for peaceful purposes.[30]

The theme of the rule of law and the development of relevant norms of space law recurs in resolution 67/113 of January 2013 together with space debris mitigation and collisions of space objects. Also contained in the Resolution is the statement that space science and technology and their applications make important contributions to economic, social and cultural development and welfare, as indicated in the resolution entitled "The Space Millennium: Vienna Declaration on Space and Human Development", adopted on 30 July 1999 by the Third United Nations Conference on the Exploration and Peaceful Uses of Outer Space (UNISPACE III), and General Assembly Resolution 59/2. The General Assembly recalled that a number of the recommendations set out in the plan of action of the Committee on the implementation of the recommendations of the Conference had been

[27] A/RES/64/86, International cooperation in the peaceful uses of outer space, 64th Session, 13 January 2010, Resolving Clause 15.

[28] *Id*, Resolving Clause 16.

[29] A/RES/66/71, International cooperation in the peaceful uses of outer space, 66th Session, 12 January 2012.

[30] *Id*, Resolving Clause 10.

implemented and that satisfactory progress was being made in implementing the outstanding recommendations through national and regional activities.[31]

The subject of global navigation satellite systems came up in General Assembly Resolution 68/75 of December 2013 where the General Assembly noted with satisfaction the continuous progress made by the International Committee on Global Navigation Satellite Systems towards achieving compatibility and interoperability among global and regional space-based positioning, navigation and timing systems and in the promotion of the use of global navigation satellite systems and their integration into national infrastructure, particularly in developing countries.[32]

In December 2014, the General Assembly adopted Resolution 69/85 where the Assembly noted *inter alia* that the Assembly was deeply convinced of the common interest of all humankind in promoting and expanding the exploration and use of outer space, as the province of all humankind, for peaceful purposes and in continuing efforts to extend to all States the benefits derived therefrom, and also of the importance of international cooperation in that field, for which the United Nations should continue to provide a focal point. The Assembly expressed its serious concern about the devastating impact of disasters, and sated stated that it was desirous of enhancing international coordination and cooperation at the global level in disaster management and emergency response through greater access to and use of space based services and geospatial information for all countries and facilitating capacity building and institutional strengthening for disaster management, in particular in developing countries.[33]

The above discussion reflects a gradual but steady progress of the United Nations toward ensuring international cooperation in the peaceful uses of outer space. Notwithstanding this emphasis on various diplomatic efforts at gathering the key players with a view to encouraging them to share technology, techniques and research on outer space activities to the benefit of all humankind equitably, and the sporadic focus on the importance of developing the norms of space law, there has been restraint on addressing the semantics of delimitation of outer space. As the ensuing discussion reflects, this has not been for want of work by the United Nations. The sheer contentious nature of the subject has only led to argumentation and, as some have termed it, a comedy of errors.

[31] A/RES/67/113, International cooperation in the peaceful uses of outer space, 67th Session, 14 January 2013, Resolving Clause 21.

[32] A/RES/68/75, International cooperation in outer space activities, 68th Session, 16 December 2013, Resolving Clause 16.

[33] A/RES/69/85, International cooperation in the peaceful uses of outer space, 69th Session, 16 December 2014, Preambular clause.

2.3 Definition and the Delimitation of Outer Space

The issue of what "outer space" is first emerged in 1966 when UN General Assembly Resolution 2222(XXI), while endorsing the Outer Space Treaty[34] requested that UNCOPUOS commence a study of questions related to outer space—an issue Zhukov in 1966 called the "altitude frontier of State sovereignty".[35] Contrary to the current trend of sweeping the issue under the carpet seemingly to avoid a protracted debate that may hamper the progress of commercial space travel and obfuscate the functionality of space transport, in the 1950s and 1960s, delimitation was considered a prerequisite for progress in the law and regulation of space transport. For instance, Sweden, with the support of other States, had, in 1958 demanded in the United States that the exact altitudinal limit be established demarcating the optimal height of air space and the minimum altitude at which outer space could be identified.[36] However, this issue was not discussed specifically at that time. The United Nations, while cognizant of the need for such a demarcation, adopted Resolutions that clearly established the untrammelled use of outer space by all countries, obviating any possibility of their extending State sovereignty in outer space.[37]

Around this time, there were views that favoured the adoption of an international treaty that established the altitude cap for air space and the height minima for outer space. A commentator distinguished at space law suggested that such a treaty should be developed on the basis of a definite altitude with due regard to the opinions of scientific, economic, political and military experts.[38] Some years later, a prominent scholar offered the view that since it appeared that at a height of approximately 100 km above sea level satellites will not be able to retain their orbit and will fall to earth, the support of a growing number of States to a demarcation between airspace and outer space at that height would be accepted.[39]

Another reason for marginalizing the distinction between air space and outer space is the functionalist approach of denying the need for such distinction, based on the argument that it is more desirable to regulate different activities of States in space rather than to identify the frontiers of air law and space law. Irrespective of the seemingly practical perception this argument creates, regrettably, this approach still leaves a lacuna in space law.

In 1977, the Legal Sub-Committee of UNCOPUOS issued a report where it was said that there were two different approaches to the issue. One view was that there

[34] Treaty on Principles Governing the Activities of States in the Exploration and Use of outer Space, Including the Moon and Other Celestial Bodies, UNOOSA, ST/SPACE/61/Rev.1. 18 I.L.M. 1434 (1979).

[35] Zhukov (1966), p. 485.

[36] *Id.* 486.

[37] See UNGA/RES/1721 (XVI) and UNFGA/RES/1962 (XVIII).

[38] Meyer (1966), pp. 221 and 222.

[39] Goedhuis (1977), p. 308.

was a need for defining outer space. The arguments supporting this view claimed that although the activities at that time carried out by States in outer space did not seem to violate the sovereignty of States, emerging new activities in outer space at lower altitudes including the anticipated launch of the space shuttle were becoming a reality and that a clear definition of outer space was necessary. The other view was that there was no urgent need for such a definition which would not be politically acceptable. It was further averred in support of this argument that no conflicts had arisen on the issue despite an increasing proliferation of outer space activities.[40] The same report cited a proposed definition of ITU of a spacecraft as "a man-made vehicle which is intended to go beyond the major part of the earth's atmosphere" without any further inquiry as to what "beyond a major part of the earth's atmosphere" meant.[41] There was also an ambitious but ambivalent definition which identified "deep space" as "space at a distance from earth equal to or greater than the distance between the earth and the moon".[42]

The view of Egypt, as contained in the Report, offered a compromise between the functionalist and spatialist approaches, claiming that both approaches were not mutually exclusive but may be considered complimentary to each other, where space activities could be defined pragmatically as they developed while the spatial approach could be regarded as the ultimate goal to be attained.[43] Germany was of the view that endeavours should be made to have a definition of outer space in the not too distant future. Iran submitted that there was an urgent need for a solution. Kuwait, Mexico and Nigeria agreed with Iran. Indonesia came up with the interesting proposition that the definition and delimitation of outer space must be based not on a particular altitude, but on the requirements of outer space technology.

Israel suggested convening an international conference on the subject on the definition and organization of outer space, while Italy reiterated a statement it had made in 1971—that the lower limits of outer space cannot be determined by scientific or technical criteria, but rather on the basis of common sense, having

---

[40] The Question of Definition and/or the Delimitation of Outer Space, Background Paper Prepared by the Secretariat, Addendum, *A/AC.105/C.2/7/Add.1*, 21 January 1977 at 4.

[41] *Id.* 6.

[42] *Ibid.* Notably, in the Report of the Legal Sub-Committee of UNCOPUOS it is recorded that Brazil's submission was that it was paradoxical to formulate juridical norms to regulate matters concerning outer space in the Legal-Sub Committee but to be unable to agree on what "outer space" meant precisely. Ecuador agreed. Chile, apparently in support, said that an agreement on the subject would put an end to an ambiguous situation which could give rise to conflicts and left in doubt the sphere of application of legal standards. Colombia made the important point that it was necessary to define the parameters of outer space in the recognition of the *sui generis* character of the geo-stationary orbital segments and an unequivocal declaration that they come under the sovereignty of countries situated on the equator. France agreed that there was a need for precise definition of outer space while Canada and the United Kingdom saw no pressing need for such definition, on the basis that no practical difficulties had arisen since the entry into force of the Outer Space Treaty in 1968. The United States maintained that unless the purpose of a definition of outer space was precisely determined, there was no use in developing a definition.

[43] *Supra* note 40 at 11 (this chapter).

taken into account the legal and political factors.[44] Poland was of the view that it was timely that space law be developed in accordance with Article 13 Paragraph 1[45] of the United Nations Charter which should lead to a legal definition in a number of fields, including the definition or delimitation of outer space. Romania underlined the importance of developing a definition of outer space, provided such definition would respect the notion of national sovereignty and access to all States to outer space for scientific research and the use of outer space for peaceful purposes. USSR was of the view that a definition of outer space should be legal, drawn up by the COPUOS Legal-Sub Committee.

A report released by UNOOSA in 2002[46] contained a submission of the then USSR, where, in 1987 in a working paper submitted to COPUOS USSR said that general agreement might be reached to the effect that any object launched into outer space could be considered as being in outer space at all stages of its flight after launch at which its altitude above sea level is 110 km or more. Furthermore, the USSR contended that space objects of States should retain the right to fly over the territory of other States at altitudes lower than 110 km above sea level for the purposes of reaching orbit around the Earth or proceeding on a flight trajectory beyond the confines of that orbit, and for the purpose of returning to earth.[47] There was no agreement on the proposed USSR text within COPUOS on this proposal. However, the conclusion reached in the COPUOS Report is encouraging. It concluded that: consensus had been reached on the referral of the issue to the Scientific and Technical Subcommittee for consideration; the preparation and updating of background papers on the question; the establishment of a working group to consider the issue on a priority basis; the consideration of issues relating to aerospace objects; the finalization of a questionnaire on possible legal issues with regard to aerospace objects; the preparation of a comprehensive analysis of the replies received to the questionnaire; the questionnaire serving as a basis for future consideration of the subject; and the preparation of a historical summary on the question of the definition and delimitation of outer space. It is a pity that the last issue is still in abeyance.

---

[44] *Supra* note 40 at 13 (this chapter). In 1975 Italy had proposed a demarcation line at approximately 90 km, such distance being the median of the values of 60 km beyond which no air activity can go and of 120 km approximately below which no space activity could be developed.

[45] Article 13(1) of the UN Charter requires the General Assembly to initiate studies and make recommendations for the purpose of: a. promoting international co-operation in the political field and encouraging the progressive development of international law and its codification; and to promote international co-operation in the economic, social, cultural, educational, and health fields, and assist in the realization of human rights and fundamental freedoms for all without distinction as to race, sex, language, or religion.

[46] Historical summary on the consideration of the question on the definition and delimitation of outer space, Report of the Secretariat, *A/AC.105/769*, 18 January 2002.

[47] *Id.* 5. See also Union of Soviet Socialist Republics: Working Paper, Approach to the solution of the problems of delimitation of air space and outer space, UNCOPUOS, Legal-Sub Committee, A/AC.105/c.2/L.121, 26 March 1979.

It is worthy of note that COPUOS, in April 2006 released a paper which outlined the contribution of Belgium to the work of a COPUOS working group on matters relating to the definition and delimitation of outer space wherein Belgium had submitted that the development of new commercial activities or services (tourist space flights, etc.) as well as the development of national efforts in lawmaking in the field of space activities, are underscoring the issue of definition and delimitation of outer space as a key issue. In this context it was Belgium's view that with regard to sovereignty of States, it is a key-concept for the distinction to be made between, on the one hand, the airspace and the rules thereto related and, on the other hand, outer space and the principles stated by the applicable international law. The conclusion was that this concept must be assessed in particular in connection with the functional approach in the definition of outer space (according to which some international space law principles apply to the space object and its crew wherever they are located).[48]

2.3.1 The Antarctic Treaty

In the context of delimitation of outer space the Antarctic Treaty[49] stands as a scholastic, if not practical analogy. The Treaty was signed at Washington in December 1959 by 12 nations.[50] The Treaty entered into force on 23 June 1961 and the signatory States came to be treated as consultative nations. There is nowhere in the Treaty the principle (as in the outer space treaty of 1967) that States cannot appropriate property. Instead, Article IV provides that nothing contained in the Treaty shall be interpreted as: a renunciation by any Contracting Party of previously asserted rights of or claims to territorial sovereignty in Antarctica. Article IV goes on to say that nor would a renunciation or diminution by any

---

[48] Contribution of Belgium to the Work of the Working Group on Agenda Item 8(a) entitled Matters Relating o the Definition and Delimitation of Outer Space, COPUOS Legal Sub-Committee, 45th Session, Vienna, 3–13 April 2006, *A/AC.105/C.2/2006/CRP.8*, 7 April 2006 at 3.

[49] Text: Cmnd, 913, Misc. no. 21 (1959); 9 *ICLQ* (1960), 475. See also http://www.nti.org/treaties-and-regimes/antarctic-treaty/. The Antarctic Treaty was followed by two treaties: The Convention for the Conservation of Antarctic Marine Living Resources, 1980; and The Convention on the Regulation of Antarctic Mineral resource Activities of 1988.

[50] Argentina, Australia, Belgium, Chile, The French Republic, Japan, New Zealand, Norway, The Union of South Africa, The Union of Soviet Socialist Republics, The United Kingdom of Great Britain and Northern Ireland, and The United States of America. Brazil, Bulgaria, China, Czech Republic, Ecuador, Finland, India, Italy, Netherlands, Peru, Poland, Republic of Korea, Spain, Sweden, Ukraine and Uruguay and Peoples' Republic of Korea had also achieved consultative status by April 2010 by acceding to the treaty. Since then, another 21 nations have acceded to the treaty. They are Australia, Belarus, Canada, Colombia, Cuba, Democratic Peoples' Republic of Korea, Denmark, Estonia, Greece, Guatemala, Hungary, Malaysia, Monaco, Pakistan, Papua New Guinea, Portugal, Romania, Slovak Republic, Switzerland, Turkey and Venezuela.

Contracting Party of any basis of claim to territorial sovereignty in Antarctica which it may have whether as a result of its activities or those of its nationals in Antarctica, or otherwise be interpreted as part of the Treaty. Furthermore, nothing can prejudice the position of any Contracting Party as regards its recognition or non recognition of any other State's right of or claim or basis of claim to territorial sovereignty in Antarctica. No acts or activities taking place while the Treaty is in force constitutes a basis for asserting, supporting or denying a claim to territorial sovereignty in Antarctica. No new claim, or enlargement of an existing claim, to territorial sovereignty can be asserted under the Treaty while the treaty is in force.

Article 1 of the Treaty provides that Antarctica will be used for peaceful purposes only. Any measures of a military nature, such as the establishment of military bases and fortifications, the carrying out of military manoeuvres, as well as the testing of any type of weapons is prohibited by this provision. The Treaty does not prevent the use of military personnel or equipment for scientific research or for any other useful purposes. In many ways the Outer Space Treaty of 1967 has similarities in the basic principles of exploring both Antarctica and outer space as the case may be for the benefit and betterment of humankind and the prohibitions in both treaties of weaponization of the territories concerned. Article III of the Antarctic Treaty provides that in order to promote international cooperation in scientific investigation in Antarctica, as provided for in Article II of the Treaty, the Contracting Parties agree that, to the greatest extent feasible and practicable: information regarding plans for scientific programs in Antarctica will be exchanged to permit maximum economy and efficiency of operations; scientific personnel will be exchanged in Antarctica between expeditions and stations; scientific observations and results from Antarctica will be exchanged and made freely available. Also the provision acknowledges that every encouragement will be given to the establishment of cooperative working relations with those Specialized Agencies of the United Nations and other international organizations having a scientific or technical interest in Antarctica.

The Treaty prohibits nuclear explosions in the territory of Antarctica. In the case of disputes between two or more Contracting Parties, Article XI provides that if any dispute were to arise between two or more of the Contracting Parties concerning the interpretation or application of the Treaty, those Contracting Parties are required to consult among themselves with a view to having the dispute resolved by negotiation, inquiry, mediation, conciliation, arbitration, judicial settlement or other peaceful means of their own choice. Any dispute not so resolved is required, with the consent, in each case, of all parties to the dispute, to be referred to the International Court of Justice for settlement; but failure to reach agreement on reference to the International Court would not absolve parties to the dispute from the responsibility of continuing to seek to resolve it by any of the various peaceful means referred to in paragraph the first part of the provision.

One of the salient features of the treaty is the liberal monitoring and inspection process, with the involvement of observers that may be put in place unilaterally by a contracting State, that keeps States informed of activity in the territory. From a jurisdictional perspective Antarctica is no man's land or *res nullius*, but where nationality would govern individual instances.

A significant corollary to the Antarctic Treaty is the Madrid Protocol of 1989 which supplements the Treaty and which may have some relevance analogically when addressing delimitation of outer space, in that it prohibits mining in Antarctica. The Protocol provides *inter alia* that activities in the Antarctic Treaty area will be planned and conducted on the basis of information sufficient to allow prior assessments of, and informed judgments about, their possible impacts on the Antarctic environment and dependent and associated ecosystems and on the value of Antarctica for the conduct of scientific research. The Protocol provides that such judgments will take account of: the scope of the activity, including its area, duration and intensity; the cumulative impacts of the activity, both by itself and in combination with other activities in the Antarctic Treaty area; and whether the activity will detrimentally affect any other activity in the Antarctic Treaty area; whether technology and procedures are available to provide for environmentally safe operations. It also addresses the need to inquire as to whether there exists the capacity to monitor key environmental parameters and ecosystem components so as to identify and provide early warning of any adverse effects of the activity and to provide for such modification of operating procedures as may be necessary in the light of the results of monitoring or increased knowledge of the Antarctic environment and dependent and associated ecosystems; and whether there exists the capacity to respond promptly and effectively to accidents, particularly those with potential environmental effects; regular and effective monitoring shall take place to allow assessment of the impacts of ongoing activities, including the verification of predicted impacts.[51]

The Protocol also calls for regular and effective monitoring to take place to facilitate early detection of the possible unforeseen effects of activities carried on both within and outside the Antarctic Treaty.

2.3.2 The Maritime Analogy

The overarching theme of "common heritage of mankind" which is pervasive in The Outer Space Treaty, The Moon Treaty of 1979,[52] and The Antarctic Treaty, also resonates in the maritime regime. The United Nations in 1970 adopted a Declaration of Principles Governing the seabed and Ocean Floor which recognized and declared that the area covered by the Declaration were the common heritage of mankind. In the Convention on the Law of the Sea (UNCLOS) of 1982, Article

[51] Protocol on Environmental Protection of the Antarctic Treaty, Signed at Madrid on October 4, 1991 and entered into force in 1998. . ., Article 3(c). See http://www.ats.aq/e/ep.htm.

[52] Agreement governing the activities of States on the Moon and Other Celestial Bodies, *UN Doc. A/RES/34/68* of 5 December 1979, Article XI.

136 provides that the area and resources of the sea (all solid, liquid or gaseous mineral resources *in situ* in the area (the seabed and ocean floor and subsoil thereof beyond the limits of national jurisdiction) at or beneath the seabed, including polymetallic nodules); are the common heritage of mankind. Article 137 follows closely behind, prescribing that no State can claim or exercise sovereignty or sovereign rights over any part of the area; or its resources, nor shall any State or natural or juridical person appropriate any part thereof. No such claim or exercise of sovereignty or sovereign rights nor such appropriation is recognized by UNCLOS. Article 137 goes on to say that all rights in the resources of the area are vested in mankind as a whole. The resources referred to in Article 136 are not subject to alienation. The minerals recovered from the area, however, may only be alienated in accordance with the relevant provisions of the Treaty and the rules, regulations and procedures of the International Seabed Authority. No State or natural or juridical person is entitled to claim, acquire or exercise rights with respect to the minerals recovered from the area.

The theme of common heritage in the outer space, Antarctic and maritime regimes is well subsumed by Shaw who says:

> However, while a *res communis* regime permits freedom of access, exploration and exploitation, a common heritage regime...would strictly regulate exploration, would establish management mechanisms and would employ the criterion of equity in distributing the benefits of such activity.[53]

Getting back to the Maritime analogy, although the main theme of common heritage of the law of the sea as recognized by UNCLOS applies to the seabed and ocean floor and subsoil thereof beyond the limits of national jurisdiction, sovereignty over the seas are specifically defined in UNCLOS. Article 2 provides that the sovereignty of a coastal State extends, beyond its land territory and internal waters and, in the case of an archipelagic State, its archipelagic waters, to an adjacent belt of sea, described as the territorial sea. This sovereignty extends to the air space over the territorial sea as well as to its bed and subsoil and sovereignty over the territorial sea is exercised subject to UNCLOS and to other rules of international law. Article 3 provides that every State has the right to establish the breadth of its territorial sea up to a limit not exceeding 12 nautical miles, measured from baselines determined in accordance with the Convention. Article 4 recognizes that the outer limit of the territorial sea is the line every point of which is at a distance from the nearest point of the baseline equal to the breadth of the territorial sea. The normal baseline for measuring the breadth of the territorial sea is the low-water line along the coast as marked on large-scale charts officially recognized by the coastal State.

There are various provisions in UNCLOS which demarcate each area of the sea. Article 33 provides that the contiguous zone may not extend beyond 24 nautical miles from the baselines from which the breadth of the territorial sea (12 miles adjacent to territory of a State) is measured. Article 55 defines the exclusive economic zone (EEZ) as an area beyond and adjacent to the territorial sea, subject

[53] Shaw (2003), p. 454.

to the specific legal regime under which the rights and jurisdiction of the coastal State and the rights and freedoms of other States are governed by the relevant provisions of UNCLOS. Within their respective EEZs all States enjoy a freedom of navigation and over flight, as well as the freedom to lay submarine cables and pipelines, and other internationally lawful uses of the sea related to these freedoms, such as those associated with the operation of ships, aircraft and submarine cables and pipelines. As per Article 57 of UNCLOS, the EEZ may extend to 200 nautical miles from the baselines from which the territorial sea is measured. With the definition of the EEZ UNCLOS conferred upon coastal States considerable sovereign rights with respect to the living and non-living resources, and "with regard to other activities for the economic exploitation and exploration of the zone, such as the production of energy from water, currents and winds".[54] Article 56, which provides that, in the EEZ, the coastal State has sovereign rights for the purpose of exploring and exploiting, conserving and managing the natural resources, whether living or non-living, of the waters superjacent to the seabed and of the seabed and its subsoil, and with regard to other activities for the economic exploitation and exploration of the zone, such as the production of energy from the water, currents and winds. It therefore sets out the general scope of the coastal States' rights, jurisdiction and duties in addition to the sovereign rights of the coastal State in the EEZ for the purpose of exploration and exploitation.

It is evident from the above discussion that some States which have submitted their views on delimitation of outer space have desired a resolution to the issue based on the fact that the increasing complexities of space travel demand some definite guidelines or criteria to go by. This approach is consistent with a view expressed in 1966, that there could well be complicated issues at international law with regard to State activities on earth and in outer space, and, with the development of new types of aircraft which are capable of flying both in airspace and outer space, the need for establishing the altitude frontier of sovereignty may become an absolute necessity.[55]

In 2013 the COPUOS Secretariat prepared a summary of information on national practices and legislation of States with regard to the definition and delimitation of outer space which contained the submission of States on delimitation.[56] Australia highlighted the fact that it is advantageous to domestic entities conducting space activities to have certainty as to the legal framework which applies to their activities.[57] Furthermore the Australian Government was of the view that international agreement on a spatial region in which international space law applies may be useful. It was suggested that such agreement should be pursued even if initial

[54] See Adi (2009), pp. 21–22.

[55] Zhukov (1966), p. 490.

[56] Summary of information on national practices and legislation of States with regard to the definition and delimitation of outer space, Note by the Secretariat A/AC.105/C.2/2013/CRP.8, 5 April 2013.

[57] *Id.* 3 and 7.

agreements do not fully resolve the delimitation of airspace and outer space and that the approach of defining a lower limit of outer space may be one way in which to achieve such initial agreement. Australia had amended its Space Activities Act of 1998 in 2001 and 2002 to recognize that a 100 km altitude had been set for practical purposes but it no means defined for Australia a delimitation point between airspace and outer space. Kazakhstan identified the same altitudinal cap, as did Australia. The United Kingdom submitted that the development of space transportation systems functioning seamlessly between airspace and outer space, relying on lift to fly through the air for part of their flight profile, will create uncertainties about the legal regime applicable to them. In particular, the distinct liability regimes applicable to each may be conflicting. There was a need to avoid hybrid solutions and a regulatory solution which provides seamless consideration and a degree of legal certainty for operators was recommended.[58] Mexico submitted that the delimitation of airspace and outer space should be settled at some point. Belgium supported the functionalist view and did not advocate any legal definition at the point it submitted its view. Netherland was of the view that no one could reasonably argue against the fact that satellites are operating in outer space, the question of the definition of outer space or the delimitation of airspace and outer space is not relevant.[59] It is submitted that this argument is somewhat tenuous and seems to say "one can recognize outer space when one experiences or sees it".

At a study conducted 2013 by the Space Law Committee of the International Law Association sub orbital flights within which the issue of necessity for delimitation of air space and outer space was addressed, the discussion had been divided among the delegates.[60] Wassenbergh has opined that "States should subject launches to an international legal regime, by harmonizing their national space legislation, beginning with agreements upon standard requirements for launches; who launches; how and what is launched, into which orbit, to do what, by whom, for whom, etc."[61] without addressing the delimitation issue. However, he goes on to say that there will be a compelling necessity to arrive at a common multilateral agreement and regulation as to who may have access to air space and outer space to operate a fair business in air space and outer space.[62] It does not seem to make logical sense to speak of access to air space and outer space without defining or in the least identifying what outer space is.

If the view of Australia and other States, which called for a multilateral regime of delimitation is to be pursued, it may make sense to take into consideration the position of the United States—that United States maintained that unless the purpose of a definition of outer space was precisely determined, there was no use in

[58] *Id* at 3.

[59] *Id.* 11.

[60] Questions on Suborbital flights for scientific missions and/or for human transportation, *A/AC.105/1039/Add.3*, 15 January 2014 at 2.

[61] Wassenbergh (1999), p. 325.

[62] *Id.* 330.

developing a definition. In this context the purpose would be to meet the increasingly complex demands of the commercial sector, particularly in the development of the aerospace plane where both airspace and outer space would be used in a long range flight, where, in instances where liability of the carrier would an issue and jurisdictional questions arise, a precise interpretation of location would be an essential consideration. Any multilateral attempt at delimitation should therefore be undertaken after a careful evaluation of both the functionalist approach and the spatialist approach with a view to covering the various complex exigencies of current and future space travel. In the ultimate analysis, such an exercise should veer from the enforcement model of the current multilateral treaty to a managerial model that is based on mutual cooperation and collective acceptance of States of the principle that individual instances involving space travel and problems arising therefrom should be solved through negotiation and analysis.

One should also bear in mind that the unstoppable progress being made in space travel would, with the effluxion of time, and the increasing involvement of new space faring nations, bring to bear the need for some rights to be accorded to States in their exploration and exploitation of outer space, without completely derogating the common heritage principle.

References

Adi M (2009) The application of the law of the sea and the convention on the Mediterranean Sea, United Nations-Nippon Foundation Fellow 2008–2009. Division for Ocean Affairs and the Law of the Sea, Office of Legal Affairs, The United Nations, New York, pp 21–22

Cheng B (1980) The legal regime of air space and outer space: the boundary problem functionalism versus spatialism: the major premises. Ann Air Space Law V:323–361

Goedhuis D (1977) Some observations on the problem of the definition and/or the delimitation of outer space. Ann Air Space Law II:287–309

Meyer A (1966) Stratftaten Weltraum. Thessaloniki, pp 221 and 222

Shaw MN (2003) International law, 5th edn. Cambridge, p 454

Wassenbergh H (1999) Access of private entities to airspace and outer space. Ann Air Space Law XXIV:311–331

Zhukov G (1966) Space flights and the problem of the altitude frontier of sovereignty. In: Yearbook of air and space law. McGill University Press, pp 485–494

Chapter 3
Competition and the Environment

3.1 Introduction

Naomi Klein correctly observes that climate change poses an expansive challenge to the industrial capitalist economic system.[1] This in turn affects competition and development as well as economic growth of States. In the aviation context, economic instruments enforced on the polluter would affect the competitive advantage airlines face, which has spawned a strong argument among developing countries, mostly those termed "emerging economies" who claim that developed countries, which are largely to be blamed for global warming with their industrialization over the nineteenth and twentieth centuries must bear more of the burden of recouping the damage. This argument may have some justification that goes back to the eighteenth century where population economist Thomas Robert Malthus drew a threefold distinction between material, immaterial and non-economic goods, the last being goods that were, although material, destitute of economic value and therefore did not form part of the political economy. To this category belonged air, water, the elasticity of steam and the pressure of the atmosphere which, being devoid of the capacity of being politically controlled could be used untrammelled and with impunity by States.[2] These goods affect the present day environmental economics and are given a value by climate change economics, giving rise to such economic instruments such as environmental taxes and other market based measures.

The subject of contention in aviation and the environment—market based measures on aircraft emissions—has been steeped in polarization between the developing countries and the developed countries where the former have vehemently protested against the imposition of an emissions trading scheme of the European Union on aircraft emissions, on the basis that their airlines, which are

[1] Klein (2014a), p. 47.

[2] Mayhew (2014), pp. 117–118.

R. Abeyratne, *Competition and Investment in Air Transport*,
DOI 10.1007/978-3-319-24372-6_3

trying to help in developing the economies of such countries would be burdened with levies on emissions to pay for pollution that has been caused by the developing nations. These developing countries sought a level playing field in which their airlines could compete without having to bear the cost of an issue that they have not caused.

ICAO's ongoing efforts at introducing a market based measure for aircraft emissions is therefore anchored upon the concept of Common but Differentiated Responsibilities (CBDR) where its member States would share responsibility conceptually on a common ground but their responsibilities in alleviating the problem of global warming caused by aircraft emissions would be different from airlines of other major polluting States. The CBDR principle has its genesis in Principle 7 of the Rio Declaration on Environment and Development otherwise called the Rio Conference or UNCED (United Nations Conference on Environment and Development) held in Rio de Janeiro in 1992. Principle 7 states:

> States shall cooperate in a spirit of global partnership to conserve, protect and restore the health and integrity of the earth's ecosystem. In view of the different contribution to global environmental degradation, States have common but differentiated responsibilities. The developed countries accept the responsibility that they bear in the international pursuit of sustainable development in view of the pressures their societies place on the global environment and of the technologies and financial resources they command.

CBDR has to be considered with Special Circumstances and Respective Capabilities (SCRC) of States when considering their air transport industries' capability of competing with more established carriers of the developed world.[3] What would be important for a level playing field for air transport in developing countries in terms of competition with the larger carriers of developing countries is to apply the CBDR and SCRC principles to per capita GDP; per capita emissions; emissions per unit GDP; population; historical emissions/contributions to climate change; and total current emissions.

Under the Emissions Trading Scheme (ETS) of the European Union, aircraft operators coming into or going out of European Community aerodromes are required to submit an annual emissions report to the competent authority. This report, which has to cover emissions pertaining to an entire journey irrespective of whether such journey commenced outside Europe or not, has to be verified by an independent and accredited verifier prior to submission. In the pre-trading period (2010–2011) airlines have to submit your first verified emissions report by 31 March 2011 and 2012. They are, however, not required to surrender emissions allowances equivalent to the operator's reported emissions for the years 2010 and 2011. The first trading period started from 1st January 2012 onwards. Starting from April 2013 aircraft operators will be required to surrender each year emission allowances that cover the verified reported data for the previous year.

Furthermore, before the start of each trading period aircraft operators are required to submit a revised monitoring plan for annual emissions. The first time

[3] For a further discussion on CBDR and SCRC see Abeyratne (2014a), pp. 100–113.

such a review takes place is before 1 January 2013. In performing the review they would have to assess whether their monitoring methodology can be changed in order to improve the quality of the reported data without leading to unreasonably high costs. The revised monitoring plan for annual emissions needs to be approved by the competent authority of your administering Member State.

The initial snag with this scheme is that a vast number of States objected to it on the ground that the European ETS was extraterritorial[4] in application. The latest report at the time of writing was that India had threatened to ban European airlines from its airspace and that 10 Chinese and Indian airlines had refused to comply with the ETS requirements, while many nations, including the United States, Russia and several prominent South American States and African States have lodged their strong protest against the application of the ETS.[5]

Marthinus van Schalkwyk, Minister of Tourism, South Africa, speaking at the 2012 Aviation & Environment Summit convened from 21 to 22 March 2012, in Geneva, referred to the ETS as an: "aggressive unilateralism" that could lead to retaliation through trade wars and distortion of competition. He requested EU to suspend the inclusion of aviation in this scheme for 2 years, encouraging multilateral dialogue and calling for a consensus under ICAO[6] in order for the aviation industry to "clean up its act."

On 2 November 2011 the Council of ICAO adopted a declaration unanimously urging the EU not to subject non-EU carriers to its ETS. This Declaration was sponsored by India and 25 other member States of the Council. The entire aviation community, including the EU—which has stated that it is willing to amend its scheme if an ICAO sponsored alternative is finalized—supports the view that market based measures should be global in application and should be developed through ICAO.

Things came to a head in the international aviation community in 2013 at the 38th Session of the ICAO Assembly where ICAO, more than 170 member States and numerous international Organizations pledged to introduce the elements of a global Market Based Measures (MBM) Scheme at the next session of the Assembly in 2016. These elements would then be developed into a global MBM by the Assembly. ICAO had been kicking the can down the road for more than a decade on this issue and the Assembly decided that it was time for concrete action. With a final reprieve from the international aviation community of another 3 years given in 2013, and as a desperate last measure, ICAO decided to adopt the *Strawman* approach—where the actual position of the MBM debate in ICAO is ignored and replaced by a distorted, exaggerated or misrepresented version of that position which is then falsified to reach a conclusion that is acceptable to all parties

[4] *Id.* 166–174. Also see Abeyratne (2008a), pp. 155–160.

[5] India Warns EU Over Airline Carbon Tax, *Air Letter*, and Thursday 24 May 2012 at 3.

[6] One of the overarching objectives of ICAO, as contained in Article 43 of the Convention is to foster the planning and development of international air transport so as to meet the needs of the peoples for safe, regular, efficient and economical air transport.

concerned.[7] Taking a scientific analogy, this approach can be related to the theory of falsification propounded by scientist and philosopher Karl Popper who stated that falsifiability is a criterion for deciding whether or not a theoretical system belongs to empirical science. The ICAO Council, at its 202nd Session tasked its Committee on Aviation Environment Protection (CAEP)[8] to conduct the first round of technical analyses on the *Strawman* of MBM, the initial construction of which was carried out by the Environment Advisory Group (EAG). The process has been going on since. This article examines the basis of the *Strawman* approach in the context of the development of an MBM scheme for aviation emissions and makes comments and offers recommendations.

At the 38th Session of the ICAO Assembly held from 24 September to 4 October 2013 the Assembly adopted Resolution A38-18 wherein the Assembly decided to develop a global MBM[9] scheme for international aviation and in this context requested the ICAO Council *inter alia* to finalize the work on the technical aspects, environmental and economic impacts and modalities of the possible options for a global MBM scheme, including on its feasibility and practicability, taking into account the need for development of international aviation, the proposal of the aviation industry and other international developments, as appropriate, and without prejudice to the negotiations under the United Nations Framework Convention on Climate Change (UNFCCC). In this regard the most important task assigned to the Council by the Assembly was to identify the major issues and problems, including for member States, and make a recommendation on a global MBM scheme that appropriately addresses them and key design elements, including a means to take into account special circumstances and respective capabilities and the mechanisms for the implementation of the scheme from 2020 as part of a basket of measures which also include technologies, operational improvements and sustainable

---

[7] A good example of a *Strawman* is seen in the Nizkor Project with the following dialogue: Prof. Jones: "The university just cut our yearly budget by $10,000."

Prof. Smith: "What are we going to do?"

Prof. Brown: "I think we should eliminate one of the teaching assistant positions. That would take care of it."

Prof. Jones: "We could reduce our scheduled raises instead."

Prof. Brown: "I can't understand why you want to bleed us dry like that, Jones." See http://www.nizkor.org/features/fallacies/straw-man.html.

[8] The Committee on Aviation Environmental Protection (CAEP) is a technical committee of the ICAO Council established in 1983. CAEP assists the Council in formulating new policies and adopting new Standards and Recommended Practices (SARPs) related to aircraft noise and emissions, and more generally to aviation environmental impact.

[9] Market-based measures refer to policy tools as well as market and economic instruments. They include: emissions trading, emission-related levies—charges and taxes, and emissions offsetting; all of which aim to contribute to the achievement of specific environmental goals, at a lower cost, and in a more flexible manner, than traditional command and control regulatory measures. Market-based measures are among the elements of a comprehensive mitigation strategy to address greenhouse gas (GHG) emissions from international aviation that are being considered by ICAO. See *ICAO Environmental Report*, 2010, Economic Issues, Chapter 4 at 2.

alternative fuels to achieve ICAO's global aspirational goals; and report the results of the work for decision by the 39th Session of the Assembly to be held in 2016.[10]

The genesis of ICAO's dilemma[11] in arriving at a solution acceptable to all its member States is essentially polemic where the stand taken by the European Union with its inclusion of aircraft engine emissions in its Emissions Trading Scheme (EUETS) drew sharp remonstrations and objections from many States.[12] The most fundamental argument against the EUETS was that it was an extraterritorial measure which was unacceptable to established tenets of international law.[13] The challenge facing ICAO and its member States involved (through its EAG) in the exercise that would deliver an acceptable MBM scheme to the Assembly in 2016 would lie in falsifying the *Strawman* which is developed based on the various options currently on the table. For example, under the *Strawman* approach taken by ICAO if one were to suggest that the best way to go is to impose a tax on fuel, the EAG would have to falsify this premise, perhaps taking into consideration arguments against it.[14] Is the best way to end taxes on fuel carried in the aircraft, as endorsed by ICAO Council Statements on Taxation, by supplementing revenue through increased embarkation taxes?[15] Would this proposition too have to be rendered unacceptable through the *Strawman* approach?[16] On the other hand, if one were to claim that a practical way to implement an MBM scheme would be to place a price on carbon,[17] this proposition will have to be falsified if it were to be eliminated from consideration. Another consideration involves carbon offsetting where one commentator says that:

> ...prospects for ICAO to agree stringent quality criteria for offsetting are poor: at worst they will accept all programmes, including voluntary carbon offsets, and at best they will merely defer to the UN's Clean Development Mechanism (CDM), including its flawed rules on proving that its emissions reduction offsets are additional to the business-as-usual

---

[10] For a detailed discussion see Abeyratne (2014b).

[11] See Abeyratne (2014c), pp. 57–74.

[12] See Abeyratne (2008b), pp. 5–9. Also see Abeyratne (2008c), pp. 155–160 where the author makes several suggestions on how the ICAO Council should proceed in its quest for a Global MBM scheme.

[13] In 2008, the European Union amended its Emissions Trading Scheme to include the aviation sector. Initially, in 2011, only flights between EU airports were to be included in the Scheme. From 2012, this was extended to all flights arriving at or departing from an EU airport. This meant that the ETS became applicable to any airline and its flights that are destined to Europe from anywhere in the world, and vice versa. Convention the Scheme would apply, for example, to a U.S. carrier's entire flight from New York to London, covering, *inter alia*, emissions released over American airspace and over other territorial airspace before entering European airspace. Detractors of ETS claim that this is an extraterritorial application of European law. See Abeyratne (2012a), pp. 7–46.

[14] See Alejandro Piera, Why taxes are not an option in addressing international civil aviation's carbon footprint, http://www.greenaironline.com/news.php?viewStory=2065. It is submitted that a proposal to impose a fuel tax is both impracticable and illogical. See Abeyratne (2012b), p. 90.

[15] *Ibid.*

[16] This proposal on taxation is both impracticable and illogical.

[17] Ahmad (2014), p. 115.

> scenario. Even so, offsets by their nature will not see emissions reductions in the aviation sector, nor is a 2020 carbon neutral growth target anywhere near in line with the global objective of limiting a temperature increase to under 2 degrees C. A failure to act by international aviation will place a greater burden on all other sectors of the economy.[18]

This proposition will also have to be debated with a view to falsifying it. It is submitted that, before the abovementioned issues are debated, the fundamental issues under the *Strawman* approach which ICAO should be addressing are that there is a real and compelling difficulty in getting all the States in the world to agree on a global MBM and that real technological solutions to the problem of emissions are absent. Another issue is that we have blown away the opportunity we had and there is nothing we can do. All these premises will have to be falsified.

3.2 The Work of ICAO

ICAO did not lose any time in springing to action. In October 2013, at the Air Transport Committee meeting held during the 200th Session of the Council, ICAO obtained consensus of the Committee (under whose purview the subject of aviation and the environment resided) that work should commence with a degree of intent, purpose and expediency on the directions of the 38th Session of the Assembly in relation to Resolution A38-18 and requests made to the Council thereby, as discussed above. To its credit, the Committee acknowledged at this meeting that the issue of a globally acceptable measure for MBMs had been discussed within the Council and Assembly for 7 years and little other than polarised disagreement on MBMs issues had resulted. Therefore, it was timely that ICAO took action to take measures towards the development of global market-based measures (MBMs) in accordance with the Resolution. Due to the political nature of the issue, the Committee decided that it should not be addressed within the ICAO Secretariat and that consideration be remanded to a new body to work on the issue. Thus, the EAG[19] was born.

3.2.1 The Environment Action Group (EAG)

It was decided that the goal of EAG was to select and report to the Council the best possible option for the global introduction of MBMs after 2020, which would meet

[18] Andrew Murphy Ending international aviation's $65 billion fuel tax exemption an essential step towards decarbonisation, http://www.greenaironline.com/news.php?viewStory=2070.

[19] The EAG consists of ICAO Council members while taking into account the principle of Equal Geographic Representation. Other Council members and non-Council State Representatives to ICAO could also participate in the work of the group. The ICAO Secretariat supports the work of the EAG. The chairman is elected by the Group in rotation of 6 months.

ICAO's Strategic Objectives relating to Safety, Environmental Protection and Sustainable Development of Air Transport, which had sufficient credibility and acceptability to be supported by an absolute majority of ICAO Member States during the 39th Session of the Assembly in 2016. With a view to keeping all ICAO member States in the loop on progress made by the EAG the Committee decided that EAG reports would also be discussed in an Environment and Climate Change Panel (to be constituted as a panel of the Air Transport Committee) which would be established on the lines of other ICAO Panels with specified terms of reference.

The regulatory basis for the establishment of the AEG and its work lies in Article 55 of the Chicago Convention[20] which lays down permissive functions of the Council.[21] The Terms of Reference of the AEG were established by the Air Transport Committee, according to which the AEG was tasked with: creating a set of selection criteria for possible options for a Global MBM based on three of ICAO's Strategic Objectives—Safety, Environmental Protection and Sustainable Development of Air Transport and their possible environmental and economic impacts, as well as practicability of global introduction and forecasted period of existence; studying the feasibility and practicability of the possible options of global MBM scheme, including its technical aspects, environmental and financial impacts to both developed and developing countries; matching possible MBM options including study of other options such as taxes, fuel or carbon charges, etc. that could also attain the objectives of MBMs for established selection criteria; integrating the concept of Common but Differentiated Responsibilities (CBDR) and Special Circumstances and Respective Capabilities (SCRC) into the global MBM scheme that would enable assistance to the developing States; reporting to Council on the result of selection in order to initiate a consultation process with ICAO Member States which will identify the major issues and problems of Member States and suggest appropriate means to address them; and finally, reviewing replies of ICAO Member States and in case of a positive result, preparing a final report for the Council which would contain an assessment of possible MBMs, their global introduction and basic provisions which can be included in the draft of a "Consolidated Statement of continuing ICAO policies and practices related to environmental protection – Climate change" for consideration of ICAO's 39th Session of the Assembly in 2016.[22]

[20] Supra, Chap. 1, note 5.

[21] Article 55 (c) prescribes that the Council may conduct research into all aspects of air transport and air navigation which are of international importance, communicate the results of its research to the contracting States, and facilitate the exchange of information between contracting States on air transport and air navigation matters.

[22] See Establishment of an Environment Advisory Group, *C-WP/14087*, 8/11/13 at 3.

3.2.2 The CAEP Process

CAEP, which is charged by the Council to analyse technical aspects of AEG's work, reviewed work carried out at the sixth meeting of the EAG which *inter alia* discussed the timing, venues and contents of MBM Global Aviation Dialogues (GLADs) to be held in the regional offices of ICAO. GLADs is mandated by Resolution A38-18 which required ICAO to organize seminars, workshops on a global scheme for international aviation participated by officials and experts of member States as well as relevant organizations in the 2014–2016 triennium.[23] Seven seminars were planned to be held across the ICAO regions in 2014 and 2015. The first four seminars were held in Mexico and Peru in April 2014, and in Cameroon and Kenya in June 2014.[24]

The subjects discussed at the GLADs are: what is an MBM, and why an MBM for international aviation?; offsetting, emissions trading, levies; future aviation emissions trend, a basket of measures, emissions gap; and past ICAO work on MBMs up to the 38th Assembly. Under progress of work since the 38th Assembly, the subjects addressed are the content of Resolution A38-18 on a global MBM; EAG process and the *Strawman* concept for a global MBM; and Roadmap of work up to the 39th Assembly. Under presentation and consultation are discussed the main design elements of *Strawman* in general terms, presenting everything EAG has considered to date and explaining the design elements in principle but without any specific details which are not consensual; offsetting mechanism; calculation of individual operator obligation; emission units criteria; monitoring, reporting and verification; and compliance cycle and duration of the scheme, etc.

CAEP has overseen the work of the AEG at the latter's meetings (EAG/7 in October 2014; and EAG/8 in November 2014) and conducted technical analyses on work conducted on the refinement of technical analyses presented to the EAG/7 meeting in October 2014, including the use of variable (dynamic) growth rates over time in basic calculation involving; analyses of a route-based approach, including the assessment of routes, using different criteria, as well as options for categorization of routes; and analyses of an accumulative approach, including the review of availability and integrity of past data.[25]

The results of the technical analyses concluded by CAEP, along with an update of its work on monitoring, reporting and verification (MRV) and emission unit criteria (EUC), were presented to the EAG/9 meeting in January 2015. At the EAG/9 meeting, a number of further technical analyses, most of which emanated from the original requests at the EAG/8 meeting, were added, and it was agreed that

[23] Resolution A38-18 Resolving Clause 19(b).

[24] Environmental Protection—recent Developments in ICAO, *C-WP/14205*, 17/10/14 at 2.

[25] CAEP established the Global MBM Task Force (GMTF) which is tackling technical aspects namely monitoring, reporting and verification (MRV) and criteria for emissions units to be eligible under the global scheme. The GMTF is also conducting some detailed quantitative analysis work delegated to it by the ICAO Council.

the CAEP would develop its work plan and schedule to address them. The new set of technical analyses was approved by the President of the Council acting under delegated authority.

The further results of some of the technical analyses were presented at the EAG/10 meeting in February 2015, which also reviewed the work plan and schedule of CAEP for the remaining ones. The EAG also agreed on criteria, upon which CAEP would undertake the task on comparison of different approaches.

There are three basic areas that came under scrutiny in the *Strawman* approach under EAG and the technical analysis of CAEP*: i.e.* considering an alternative to an MBM such as a levy, tax or charge; basing allocation of emissions on historical growth patterns; and allocating responsibility based on groups of flight routes, for example, defined as 'mature' or 'developing'. The CAEP task force has concentrated on technical work involving monitoring, reporting and verification—developing the detailed modalities for an MRV regime that ensures robust accounting of aviation CO2 emissions under a global MBM; sources of CO2 mitigation—developing the eligibility criteria of emissions units (allowances, offsets) for use within an MBM; and technical and quantitative analysis of certain aspects of the *Strawman* proposal.

During the CAEP/9 meeting held in Montreal in February 2013 it was agreed that an ICAO CAEP Aviation Environmental Impacts Seminar would be organised to study the state of scientific knowledge, focusing on four subjects: (1) Aviation Impacts on Climate; (2) Climate Impacts on Aviation; (3) Aviation Impacts on Air Quality; and (4) Aviation Noise Impacts. The seminar was held in Washington, DC, United States from 10 to 12 February 2015 and attended by approximately 100 international scientific experts. The seminar provided the means by which the CAEP Impacts and Science Group (ISG) will begin drafting White Papers on the consensus state of science, with the aim of delivering the finalised White Papers to the CAEP/10 meeting in February 2016.

3.3 The Strawman Approach

The main characteristic of the "*Strawman*" is that it is a tentative draft proposal which can be taken apart as more information becomes available.[26] Therefore a *Strawman* helps one to make a start in a project and build on propositions as one goes along and gives a tangible starting tool. Often the start up information is

---

[26] A *Strawman* can take any form. It can be the outline of a new strategy, revenue and expense projections, a future org chart, or even the start of a case statement. It's something that isn't considered definitive, but it's meant to create a reaction and shape discussion and future action. The final outcome may look nothing like the *Strawman*, but the *Strawman* helps you to get there. A *Strawman* helps to ensure that every idea is uncovered, vetted, and molded into something that will translate into action—See more at: http://www.frontrangesource.com/iron-man-strength-strawman/#sthash.GzPdRBna.dpuf.

incomplete and is developed with the techniques of research, testing and criticism. The entire process is based on falsification of existing propositions, which is calculated to assist in reaching a final viable proposition. As already stated, the scientific basis for the *Strawman* approach was propounded by Karl Popper. Popper maintained that a theory is falsifiable if there is at least one non-empty class of homotypic basic statements (of the same type or structure) which are forbidden by it.[27] Popper was of the view that empirical (scientific) and trans-empirical (non scientific) questions and answers could replace a meaning proposition which would lead to testability, verifiability, falsifiability and refutability. The decisive feature for Popper's theory was the easy verifiability whether a law, principle or proposition that is formulated could be refuted by evidence.[28] Thus the *Strawman* becomes old conjecture that should be subjected to the most severe and searching criticism. The growth of knowledge and reasoning therefore proceeds through the elimination of error and refutation of hypotheses that are either logically inconsistent or involve empirically refutable consequences.

For instance, in the instance of developing a global MBM scheme, one could start by creating a draft proposal to adopt the text of a multilateral treaty on MBMs for aviation emissions. In that proposal the author can, based on knowledge, initial judgment and experience, suggest advantages for such a proposal, such as the creating of a legal process; establishment of a structured CBDR process and SCRC justification. The homotypic statements which would militate against considering a treaty could well be that the adoption and entering into force of a multilateral treaty would be a cumbersome and tedious process (which militates against a final fix of a global MBM for 2020 that is fully implementable) and that it may never be subscribed by the sufficient number of ratifications, given the contentious nature of the subject.

The first step in the *Strawman* process would be an analysis of the proposal on the basis that it is merely a *Strawman* at that point. The next measure, which is extremely important, would be to analyze the weak points in the proposal and reasons as to why it would not work. The final step would be to draft a new proposal and keep repeating the process until a viable solution is reached by the team. Some of the criteria that could be applied to the *Strawman* proposal are in the following questions: Is it the right conclusion?; does it have context hierarchy (is it constructed in the right order?)? do definitions oscillate (are the propositions in the proposal constructed in the right order?); Are elements in the proposal concretized?; is it inductive; is the argument broken up into steps: are some objections being blatantly ignored?; and is the proposition properly integrated?[29]

ICAO sets out the key elements of a global MBM as a scope; an objective; a baseline; a common approach for all participants to contribute to the attainment of

[27] Popper (2002), p. 95.

[28] *The New Encyclopaedia Britannica*, Volume 25, Macropaedia, 15th Edition, 1990 at 642.

[29] See *Understanding Objectivism: A Guide to Learning Ayn Rand's Philosophy*, Lectures by Leonard Peikoff (Michael S. Berliner ed.) New American Library: 2012 at 112–113.

the objective; monitoring, reporting and verifying emissions; environmental integrity of emissions units; registry; compliance cycle; enforcement mechanism.[30] All these key elements have to be paired off against a *Strawman* proposal for validation, together with industry perspectives, which call for a global MBM: not to be used to raise general revenue or suppress demand for air travel; to maximise environmental integrity and be cost effective; minimize competition distortion; and be easy to implement in a cost effective manner.

3.3.1 Application of the Strawman *Approach to MBMs*

At the 38th Session of the ICAO Assembly, ICAO presented three options for a global MBM. The first option is *global mandatory offsetting* where offsetting operates through the creation of emissions units which quantify the reductions achieved. These emissions units, which would generally be created outside the international aviation sector, can be bought, sold or traded. A global mandatory offsetting scheme for international aviation would require participants to acquire emissions units to offset CO2 above an agreed target. Emissions units would need to conform to agreed eligibility criteria to ensure adequacy of emissions reductions. No specific aviation allowances or revenues would be created under this scheme. The second option is *global mandatory offsetting with revenue* which is global mandatory offsetting complemented by a revenue generation mechanism would generally function the same way as the mandatory offsetting scheme. A key difference would be that in addition to offsetting, revenue would be generated by applying a fee to each tonne of carbon, for instance, through a transaction fee. The revenue would be used for agreed purposes, such as climate change mitigation or providing support to developing States to reduce GHG emissions. The third option is a *global emissions trading scheme (ETS)* which would use a cap-and-trade approach, where total international aviation emissions are capped at an agreed level for a specified compliance period. Specific aviation allowances (one allowance is equivalent to one tonne of CO2) would be created under this scheme for all the emissions under the cap within the international aviation sector. These allowances would then be distributed for free, or auctioned, to participants using an agreed method. At the end of each compliance period, participants would need to surrender allowances, or other emission units, equal to the emissions generated during that period, including those above their allocation.[31] This having been said, ICAO should, for purposes of completion also consider other options such as taxes,

[30] *Global Aviation Dialogues (GLADs) on Market-Based Measures to address Climate Change: Key Elements of an Aviation MBM, Environment*, Air Transport Bureau, International Civil Aviation Organization (ICAO), Preliminary Version, 15 March 2015.

[31] For a discussion of these objectives, See *Aviation and Climate Change: In Search of a Global Market Based Measure*, Supra note 11 at 92–95 (this chapter).

charges and other levies, if only to falsify the propositions within ICAO's mandate. With all its perceived shortcomings, the possibility of a multilateral treaty should also be on the table for falsification and rebuttal, particularly considering the fact that multilateral agreement has been reached in international treaties such as the Ozone Depletion Convention and the Nuclear Disarmament Treaty[32] which can be treated as analogous to the MBM context.

The abovementioned three options are being subject to a quantitative impact analysis by the technical task force of CAEP. It is submitted that prior to this measure it would be advisable to falsify (or attempt to falsify) the basic premise of practical applicability of each of these options, taking into consideration the political realities of the overall issue of introducing a global MBM that would be acceptable to all of ICAO member States.

3.3.2 Strawman *Considerations for ICAO*

Strawman is not only a proposal or proposition that should be knocked down but it is also a requirement that certain conditions are considered in a methodology that will lead to a final conclusion. In this context ICAO has to consider the basic guidelines provided by Assembly Resolution A38-18 which are: MBMs should support sustainable development of the international aviation sector; MBMs should support the mitigation of GHG emissions from international aviation; MBMs should contribute towards achieving global aspirational goals; MBMs should be transparent and administratively simple; MBMs should be cost-effective; MBMs should not be duplicative and international aviation CO2 emissions should be accounted for only once; MBMs should minimize carbon leakage and market distortions; MBMs should ensure the fair treatment of the international aviation sector in relation to other sectors; MBMs should recognize past and future achievements and investments in aviation fuel efficiency and in other measures to reduce aviation emissions; MBMs should not impose inappropriate economic burden on international aviation; MBMs should facilitate appropriate access to all carbon markets; MBMs should be assessed in relation to various measures on the basis of performance measured in terms of CO2 emissions reductions or avoidance,

[32] The Vienna Convention on the Protection of the Ozone Layer, at http://ozone.unep.org/Publications/VC_Handbook/Section_1_The_Vienna_Convention/index.shtml. Also The Montreal Protocol on the Depletion of the Ozone Layer, at http://ozone.unep.org/new_site/en/montreal_protocol.php. The Nuclear Non-Proliferation Treaty (NPT) was an agreement signed in 1968 by several of the major nuclear and non-nuclear powers that pledged their cooperation in stemming the spread of nuclear technology. Although the NPT did not ultimately prevent nuclear proliferation, in the context of the Cold War arms race and mounting international concern about the consequences of nuclear war, the treaty was a major success for advocates of arms control because it set a precedent for international cooperation between nuclear and non-nuclear states to prevent proliferation. See https://history.state.gov/milestones/1961-1968/npt.

where appropriate; MBMs should include *de minimis* provisions[33]; where revenues are generated from MBMs, it is strongly recommended that they should be applied in the first instance to mitigating the environmental impact of aircraft engine emissions, including mitigation and adaptation, as well as assistance to and support for developing States; and, where emissions reductions are achieved through MBMs, they should be identified in States' emissions reporting.

One of the considerations worth looking into in executing the *Strawman* approach is the compellingly logical suggestion in a Report[34] released by Oxford University in late 2013 on governance for the future in the context of climate change and its effects in the long term. The Oxford report suggests that in the long term States should form a coalition which the university called a C20–C30–C40 Coalition to counteract climate change. The coalition, according to the suggestion of the university, would be made up of G20 countries, 30 companies, and 40 cities. This coalition formula could be the initial formation of the *Strawman* approach. This might be the first step ICAO might wish to consider taking.

The ICAO Council has to proceed with the *Strawman* approach in stages. The fundamental consideration is that it will be the Assembly that will develop an MBM scheme based on input provided by the Council. Therefore, unlike in the usual circumstances, where the Assembly directs the Council to develop a scheme which the Assembly endorses, the Council will have to explain fully the process it followed; the propositions it considered and discarded, and justify the criteria in recommending a suitable framework; process and final recommendation for a global MBM. This process would need the scheduling of extra time at the Assembly, and prior notification of details of the various steps taken by the Council in the *Strawman* approach. Rejection of the original *Strawman* and acceptance of validated alternatives would have to be explained in detail to ICAO member States. This would in turn make it necessary for credible studies that the Council, and indeed the Secretariat of ICAO should carry out on critical issues involved including but not limited to CBDR and SCRC.

SCRC would be a pivotal and critical feature in these studies so that the Council can ultimately "sell" its product elements that would eventually be developed into an MBM by the Assembly. It goes without saying that the Council will have to conduct consultations with member States and international Organizations (which means the Secretariat will have to do the groundwork). The Secretary General would be the person responsible in ensuring that the background work is done by his Secretariat with informed competence, knowledge and diligence. Such a task would involve not only knowledge but creativity and insight. Insight is critical which is much more valuable than perceived knowledge.

---

[33] The *de minimis* threshold would exempt States which had less than 1 % of the total international revenue tonne kilometres, from implementation of MBMs.

[34] Oxford Martin Commission for Future Generations, http://oneworld.org/2013/10/15/report-calls-for-action-on-climate-poverty-governance-and-disease/.

One of the key *Strawman* is the politically motivated and driven *per capita* approach which should be considered against the *status quo* approach. The *per capita* approach claims that wealthy nations should not be given an entitlement or "blank cheque" to their existing rights. This argument based on fairness is plausible to the extent that a welfarist approach would result in maximizing the pie and distributing it equally. Developing States tend to have larger populations which would do well with the welfare or per capita principle.

Any global MBM which has no compelling compliance mechanism and clear measures and implementing process would be destitute of effect. This mechanism will have to be legislative in nature or in the least have quasi legislative features and attributes, such as Standards in an Annex to the Chicago Convention. Ideally, these principles should be incorporated in the legal system of a participating State. The rules would have to be precise and offer a fair deal to every State concerned. The Parties themselves would have to agree on authentic interpretations of the provisions of the scheme. There are precedents for this process. For example, since 1979 the Convention on Long-range Trans-boundary Air pollution (LRTAP) which has addressed some of the major environmental problems of the UNEP[35] through scientific collaboration and policy negotiation, and which has been extended by eight Protocols that identify specific measures to be taken by Parties to cut their emissions of air pollutants, has adopted "common understandings" to define specific terminology of the Sulphur Protocol of 1985. In the 1983 International Undertaking on Plant Genetic Resources, The Food and Agriculture Organization—a specialized agency of the United Nations the same as ICAO, has done the same thing to clarify and establish universally acceptable interpretations.

Key States must effectively participate in the law making process of the scheme and ensure participation in a global MBM. The MBM should come with clear and stringent provisions for violations. Polarization, as was seen in the significant opposition to the European Emissions Trading Scheme by emerging economies and economic giants such as the United States and China would only continue to stultify the process. The end result of such a scenario would be individual legislation by States which would be an unfortunate failure of ICAO.

Finally, ICAO should give some consideration to the deployment of the global MBM and the use to which it could be put. Most ICAO member States, if not all, would not be averse to investment in renewable energy. There is compelling evidence presented in a major study conducted in 2012 in the United States' Department of Energy's National Renewable Energy Laboratory that wind, solar and other currently available technologies could meet 80 % of Americans'

[35] The United Nations Environment Programme (UNEP) established in 1972, is the voice for the environment within the United Nations system. UNEP acts as a catalyst, advocate, educator and facilitator to promote the wise use and sustainable development of the global environment. UNEP work encompasses; Assessing global, regional and national environmental conditions and trends; Developing international and national environmental instruments; and Strengthening institutions for the wise management of the environment. See http://www.unep.org/About/.

electricity needs.[36] It is claimed that, technically speaking, it is entirely possible to rapidly switch our energy systems to 100 % renewables globally by as early as 2030.[37] One would not dispute that any resources generated by a global MBM could be considered a good investment in research conducted for renewables.

References

Abeyratne R (2008a) The new emissions trading scheme: airlines – is it extraterritorial? Environ Policy Law 38(3):155–160

Abeyratne R (2008b) The authority of the European Union to unilaterally impose an emissions trading scheme. Air Space Lawyer 21(4):5–9

Abeyratne R (2008c) The new emissions trading scheme: airlines – is it extraterritorial? Environ Policy Law 38(2):155–160

Abeyratne R (2012a) Aircraft emissions – in search of a global framework for market based measures. Issues Aviat Law Policy 12(1):7–46

Abeyratne R (2012b) Aeronomics and law: fixing anomalies. Springer, Heidelberg, p 90

Abeyratne R (2014a) Aviation and climate change: in search of a global market based measure. Springer, Heidelberg, pp 100–113

Abeyratne R (2014b) Aviation and climate change: in search of a global market based measure. Springer, July ISBN: 978-3-319-08442-8 (Print) 978-3-319-08443-5 (Online)

Abeyratne R (2014c) Aviation and climate change: a compelling legal challenge for a toothless tiger? J Int Comp Law 1:57–74

Ahmad MT (2014) Environmental effectiveness of ICAO's basket of mitigation measures to arrest emissions from international civil aviation. Ann Air Space Law XXXIX:75–128

Hand MM et al (2012) Renewable electrical futures study, vol 1: exploration of high penetration renewable electricity futures. National Renewable Energy Laboratory, pp xvii–xviii

Klein N (2014a) This changes everything. Knopf, Canada, p 47

Klein N (2014b) This changes everything, capitalism vs. the climate. Knopf, Canada, p 101

Mayhew RJ (2014) Malthus: the life and legacies of an untimely prophet. Harvard University Press, Cambridge, pp 117–118

Popper K (2002) The logic of scientific discovery. Routledge, London, p 95

[36] Hand et al. (2012), pp. xvii–xviii.

[37] Klein (2014b), p. 101.

Chapter 4
Competition and Safety

4.1 Introduction

Competition and safety are integrally interlinked, in that an airline's safety record would act in either promoting the services of an airline or diminishing the airline's competitive advantage over other carriers which maintain a good safety record. The events of 2014 where aircraft either just disappeared or were shot down, and in one case could not survive adverse weather conditions and just crashed into the sea, have spurred the international community to initiate various measures such as the protection of safety information and warnings with regard to flights over conflict zones.

4.2 Impact of Competition on Safety

OECD in its Report *Emerging Risks in the 21st Century – An Agenda for Action* says that:

> The openness and connectedness of systems and the mobility of people, goods, services, technology and information increase the number of potential interactions that can generate or influence a hazard. Risks become more complex. At the same time there is greater awareness of the complexity of the world itself (*e.g.* of natural or social processes), and of the need to better account for that complexity when considering risk issues... Regulatory change and the development of transport, trade and information systems mean that many activities depend on the interaction of a variety of actors within networks, often at a global scale. With regard to risk this is positive, to the extent that information gathering and processing are facilitated, as are contacting victims and organising help. But connectedness also multiplies the channels through which negative consequences can propagate.[1]

[1] OECD:2013, Paris at 12.

R. Abeyratne, *Competition and Investment in Air Transport*,
DOI 10.1007/978-3-319-24372-6_4

Liberalization of air transport is irreversible. In view of the strong forecast for sustained growth of air transport, a change in the focus of accident prevention efforts is inevitable and necessary. In addition to a robust framework of regulatory requirements and approved procedures, a more proactive approach to ensure that safety is effectively and responsibly managed is absolutely necessary. Notwithstanding the fact that international civil aviation is a very safe mode of transportation, there are many challenges that need to be addressed in order to achieve a further reduction in the accident rate. Such a reduction is required to prevent the number of fatalities and accidents from rising as traffic increases, which, if not achieved, could undermine public confidence in the safety of the global air transport system.

The issue of the potential impact of liberalization and consequent competition between carriers on safety and security and their interrelationship dates back some years. It was the subject of discussion at the two most recent ICAO air transport conferences held in 1994 and 2003. The discussions at both conferences and conclusions flowing therefrom reflected a consensus that liberalization is a desirable goal that should be pursued by each State at its own choice and own pace. At the same time, the conferences were unequivocal that safety and security must remain of paramount importance, irrespective of any change in regulatory or commercial arrangements arising from demand for air services and competition. The widespread view was that, as liberalization spreads, and competition among carriers pervade the international market, there continues to be a need to address existing as well as potential concerns over its implications on safety and security. Therefore, States are faced with the challenge of maintaining uncompromising standards of safety whilst maximizing the benefits of economic liberalization.

In the air transport field, liberalized policies (e.g. on market access, airline designation, capacity, pricing, and commercial opportunities) drive the development of the industry, bringing about many economic benefits for States, the industry and consumers, such as growth in traffic (both in terms of passenger/cargo traffic and aircraft movements). Competition as a by-product of liberalization has spawned multiple air carriers (including low-cost carriers) who have entered the market offering competitive pricing which the legacy carriers find difficult to match. In addition, increased service options and development of travel and tourism have created jobs in the aviation and allied industries, making liberalization indispensable to the air transport industry. Although there is no indication that safety standards in many liberalized markets have not been maintained, nonetheless, it is also clear that the resulting growth in air transport activity and complex commercial arrangements stemming from increased business and operating practices could put additional pressure on the State in terms of its capacity in safety regulation. A State is required to provide safety oversight both to its own aircraft operators and those foreign operators that operate in its airspace. This puts an enormous burden on a State just to cope with the consequences of liberalization requiring that State to have the necessary legal, regulatory and organizational infrastructure and human and financial resources in place in order to perform the required safety/security regulatory functions.

Many of ICAO's 191 member States are already facing problems with respect to safety oversight. A glaring fact emerging from safety audits conducted by ICAO on States is that the findings of the initial safety oversight audit conducted by ICAO relating to Annex 1—*Personnel Licensing*, Annex 6—*Operation of Aircraft* and Annex 8—*Airworthiness of Aircraft*, indicated that of the 181 Contracting States that were audited between March 1999 and July 2004, considerable numbers of States had deficiencies in respect of a number of requirements under these Annexes. Furthermore, audit follow-up missions have revealed that in many cases, significant deficiencies identified during the initial audits remain. Therefore, the ineluctable conclusion is that, when considering liberalization, States should be concerned not merely of the economic benefits that would result but also its potential impact on safety regulation. States must ensure their continued capacity to meet those requirements so that, as prescribed in the Chicago Convention, civil aviation develops in a safe and orderly manner.

Over the past 70 years, significant improvements have been made in the safety of the international aviation system. The decrease in fatalities is tremendous as seen in the figures of 1945 where a rate of 4.48 passenger fatalities per 100 million passenger miles was recorded. In 1995 the rate had dropped to 0.04. Thus, over a period of 50 years, the risk of fatalities to the flying public was reduced by a factor of 100. More recently the situation had improved even further. ICAO recorded in 2013 that Compared to 2011, the number of accidents 1 in 2012 decreased by 21 % and the number of fatalities decreased by 10 % making 2012 the safest year with regard to fatalities since 2004. Due to the reduction in accidents, combined with the increase in departures, the global accident rate involving scheduled commercial operations for 2012 has decreased significantly to 3.2 accidents per million departures.[2]

It is unquestionable that accident numbers are not evenly distributed through the regions. To obtain a more complete picture, regional rates have been established using 5-year averages due to the volatility inherent in the annual accident rates caused by the small amount of traffic in regions such as Africa and the Middle East. In these regions, annual accident rates depend heavily on the circumstances of a particular year.

4.3 Safety Information

The Second High Level Safety Conference of the International Civil Aviation Organization was held in Montreal from 2 to 5 February 2015. The main focus of discussion was on the protection of safety information pursuant to aircraft accidents and incidents and the establishment in ICAO of an information exchange prototype through an ICAO webpage containing input given to ICAO by authorized officials

[2] *State of Global Aviation Safety*, ICAO:2013 at 4.

of member States. These input would pertain to operational restrictions and prohibitions issued in response to risks associated with conflict zones. The conference strongly supported the development of such a capability and suggested that a simple web-based system should be developed and implemented without delay. With regard to safety information, the protection of which could prevent persons giving such information from being criminalized, the Conference agreed that an urgent need existed for ICAO to establish working groups towards developing appropriate Standards and Recommended Practices (SARPs) that could protect safety information. Although the common theme of the Conference was a call for urgent ICAO action on these two fronts, there was some confusion where, at an information session prior to the Conference, a reputed International Organization made the startling suggestion that States should be required to make ICAO SARPS on these issues mandatory components of their internal laws. One State strongly suggested that, while it was appropriate to set up a globalized or a centralized information sharing facility as regards the use of airspace and potential threats to flights, a thorough legal assessment of all those provisions regulating the activities of such a system was necessary prior to action being taken. This article addresses the discussions at the Conference and inquires into their legal nuances.

At the First ICAO High Level Safety Conference, convened in Montréal from 29 March to 1 April 2010[3] delegates attending the conference agreed consensually that sharing and transparency of safety information was vital. A theme repeated in many of the papers discussed was the need for a free flow of information that could be used to assess safety risks and develop appropriate responses. It was also recognized that a critical factor in relation to information sharing was that there must be a clear understanding of how and to whom safety information would be disseminated and also how such information would be used. The Conference further recognized the critical importance of examination of legal aspects of information sharing in pursuance of these objectives.

4.3.1 Considerations of the Conference

Some of the key considerations of the Conference were that ICAO should coordinate the integration of safety information provided by the international community

[3] The ICAO Council, at the Fourth Meeting of its 188th Session convened on 26 October 2009, approved the convening of the First High Level Safety Conference at the level of Director General, Civil Aviation of member States. See C-DEC 188/4, 27/10/09 at 2. The purpose of the Conference was to bring together the appropriate level of senior management within States' civil aviation authorities to build consensus, obtain commitments and formulate recommendations deemed necessary for the effective and efficient progress of key safety activities by ICAO and Contracting States. The agenda of the Conference had three main themes: (a) the foundations for global aviation safety; (b) towards the proactive management of safety; and, (c) harmonization of rules and processes to address other safety issues.

and the dissemination of safety intelligence, as appropriate; that ICAO should convene a group of experts to address the issue of the definition and harmonization of safety metrics, associated data requirements and analysis processes; that ICAO, States and all interested parties should ensure that measures they take to improve safety, including those related to foreign operators are based on consistent safety indicators; and that ICAO should develop a code of conduct on the sharing of safety information to ensure that such information is used solely to improve aviation safety and not for inappropriate purposes, including for the purpose of gaining economic advantage.[4]

On the subject of protection of safety information the Conference recognized the crucial importance of trust in the protection of safety data from inappropriate use by aviation organizations and to distinguish where and when safety data/safety information/safety intelligence needs to be protected from misuse. Delegates recognized that the development of best practices for the protection of safety information was important if the free flow of information were to be sustained. The Conference was of the view that voluntary data be imperatively offered protection and that the protection be commensurate with the data reported. It was further suggested that recommendations from accident and incident investigations be followed up and the effectiveness of their implementation be monitored and confirmed.[5]

4.3.2 Recommendations

Two key Recommendations of the Conference were that ICAO should convene a group of experts to determine harmonized safety metrics, associated data requirements and processes to enable integrated safety analyses and to ensure consistent development of related safety measures, and that ICAO should facilitate the integration of safety data and information from various sources as well as the dissemination of related analysis output, with the objective to ensure that such information is disseminated for appropriate purposes.[6]

These discussions and Recommendations were seemingly a good lead to the Second ICAO High Level Safety Conference, convened on 2–5 February 2015 in the aftermath of the disappearance of Malaysian Airlines Flight MH 370 and the destruction of Malaysian Airlines Flight MH 17 in 2014.[7] The Conference was presented with a Declaration to be adopted at the end of its session which *inter alia* called upon ICAO to adopt new and enhanced provisions on the protection of safety

[4] See REPORT OF THE HIGH-LEVEL SAFETY CONFERENCE 2010 Montréal, 29 March–1 April 2010, *Doc 9935*, *HLSC 2010*, pp. 1–6 at Paragraph 37.

[5] *Id.* paragraph 39.

[6] REPORT OF THE HIGH-LEVEL SAFETY CONFERENCE (HLSC 2010), *AN-WP/8496*, 16/04/10, Appendix A, at A-4.

[7] See Abeyratne (2014), pp. 329–342. Also see Abeyratne (2015), pp. 2–51.

management information as well as accident and incident records and support States in their implementation; and develop a global information sharing framework to collect and share harmonized safety information and provide the means to adequately protect the resulting safety information.[8]

4.4 Discussions at the Second High Level Safety Conference

4.4.1 Protection of Safety Information

At the Conference, the ICAO Secretariat brought to bear the need to supplement Annex 19 to the Convention on International Civil Aviation[9] by proposing that a balance be stricken between the need for the protection of safety management information and the need for the proper administration of justice. There was also a proposal to establish parameters to ensure that safety management information is available to be used for its intended purposes by determining the levels of protection to be afforded to safety management information appropriate to specific circumstances; and providing necessary flexibility for Contracting States in determining the competent authority to make decisions regarding the disclosure of safety management information for use in judicial, administrative and disciplinary proceedings, as well as to the public.[10]

The United States and Brazil submitted to the Conference that one of the key issues to be addressed by the aviation community involved protection of safety data and information with a view to allowing data that would facilitate the conduct of safety management in areas of certification, operations, analysis and investigation, in an environment where safety reporting can occur free from harm of inappropriate use. While recognizing the need to conduct aircraft accident investigations thoroughly with all relevant information, the two States brought to the attention of the Conference the importance of protecting the sources and use of safety data and information with a view to ensuring that information is used only for maintaining or improving safety, rather than taking punitive action against individuals or organizations providing this information.[11]

The commonality between the contention of the ICAO Secretariat on the one hand, and the United States and Brazil on the other is that both advocate a harmonious blending between the elements of freedom to provide information

[8] Montreal Declaration on Planning for Aviation Safety Improvement *HLSC/15-WP/108*, 5/2/15 at 3.

[9] *Supra* Chap. 1, note 5. Annex 19 to the Chicago Convention pertains to safety management systems.

[10] Balancing the Use and Protection of Safety Information, *HLSC/15-WP/4*, 8/10/14, p. 3 at paragraph 2.1.3.

[11] Safety Information Protection (sip), *HLSC/15-WP/25*, 20/11/14, at p. 3, paragraph 3.1.

and compulsion as well as the protection of the individual providing the information from punitive sanction.

In a paper presented by Latvia on behalf of the European Union and its Member States and the other Member States of the European Civil Aviation Conference and also by EUROCONTROL,[12] the Conference was requested to support "enhanced provisions on the protection of accidents and incident investigation records and the protection of video or audio recordings of cockpit voice recorders or airborne image recorders in Annexes 6 (*Operation of Aircraft*) and 13 (*Aircraft Accident and Incident Investigation*) to the Chicago Convention", on the basis that in order for safety hazards to be identified and risks to be determined, there must be a continuous flow of safety information. The paper pointed out that for a proper risk assessment to be conducted a variety of sources must be tapped and that exchange of information between partners such as regulated stakeholders and aviation public authorities was crucial. In this context the sources of information must have sufficient confidence in the protection offered by the different systems available to report safety occurrences, including those involving their own mistakes and the exchange of safety information must be accompanied by relevant and adequate limitations regarding the use of such information.[13]

Several States[14] complained that the vulnerability of safety information lay in the fact that such information had been used for disciplinary and compliance purposes, where in certain instances, the information provided by those involved in incidents and accidents had been used to prefer criminal charges against them. It was further claimed that this trend portends a negative impact on the implementation of aviation safety management systems, adversely affecting the prevention of incidents and accidents.[15] These States concluded that States, in order to enact appropriate legislation to protect safety information, should be provided with familiarization of the scope of safety information protection. ICAO was requested to incorporate definition of data, information, intelligence information *et al.* so that criteria could be established that would determine the scope of protection.[16]

As a compromise to the dichotomy between protection of the information source and justice, as well as to ensure the free flow of information, The Civil Air

---

[12] Protection of Safety Information, HLSC/15-WP/38 15/12/14, at p. 1.

[13] *Id.* p. 3 at paragraph 3.1.

[14] Peru, Argentina, Brazil, Colombia, Chile, Ecuador, Panama, Paraguay, Uruguay, Bolivia and Venezuela.

[15] Challenges for the Implementation of Safety Information Protection Measures, *HLSC/15*-WP/46 23/12/14, at p. 1, Paragraph 1.2.

[16] *Id*, p. 3 at Paragraphs 3.1 and 3.2.

Navigation Services Organization (CANSO)[17] presented a *Just Culture* which contained the basic principle that the open reporting and discussion of safety issues and mistakes was essential in the understanding that one must be individually held to account for one's actions and that everyone is responsible for acting safely in a manner which is commensurate with one's training, experience, and the professional standards expected in his/her job.[18] These Just Culture Guidelines are based on the premise that staff safety reports are one of the most valuable sources of information for learning safety lessons and therefore feel the need to foster a culture in which staff feel secure that the organization will treat them justly and fairly when they report. The elements of the CANSO *Just Culture* are: staff responsibility; organizational responsibility; safety reports; protection and support; and zero tolerance for unacceptable behaviour.

Airports Council International (ACI)[19] reiterated the principle that the primary purpose of protecting safety information from inappropriate use is to ensure its continued availability so that proper and timely preventive, corrective or remedial actions can be implemented and aviation safety maintained or improved.[20]

The Conference concluded *inter alia* that ICAO should make meaningful progress towards the adoption of new and enhanced provisions on the protection of safety management information as well as accident and incident records, while ensuring maturity, consistency and clarity on the proposals. The Conference requested that States undertake the necessary legal adjustments to efficiently implement new and enhanced protective frameworks to facilitate safety management and accident investigation activities; and ICAO should support States in implementing new and enhanced provisions through a strategy comprised of supporting guidance material, tools and seminars tailored to the needs of each region aiming at building trust, cooperation and a common understanding among aviation safety professionals, accident investigation authorities, regulators, law enforcement officers and the judiciary in the context of an open reporting culture.

---

[17] The Civil Air Navigation Services Organisation—is the global voice of air traffic management (ATM) worldwide. CANSO Members support over 85 % of world air traffic. Members share information and develop new policies, with the ultimate aim of improving air navigation services (ANS) on the ground and in the air. CANSO represents its Members' views to a wide range of aviation stakeholders, including the International Civil Aviation Organization, where it has official Observer status. CANSO has an extensive network of Associate Members drawn from across the aviation industry.

[18] CANSO Guidelines on Just Culture, HLSC/15-WP/54 *Appendix*, at p. 1.

[19] Airports Council International (ACI) is the only global trade representative of the world's airports. Established in 1991, ACI represents airports interests with Governments and international organizations such as ICAO, develops standards, policies and recommended practices for airports, and provides information and training opportunities to raise standards around the world.

[20] ACI Views on the Secretariat Working Papers, *HLSC/15-WP/88* 26/1/15 (Revised), at p. 2.

4.4.2 Legal Issues Involved

There are two broad issues involved here. The first is the somewhat naive and impracticable suggestion made by some that ICAO should adopt SARPs and ensure that States incorporate them in their legislation. ICAO has neither the standing nor authority to do this. Besides, Article 38 of the Chicago Convention[21] makes SARPs non-prescriptive. The argument adduced is that "ICAO" treaties when ratified by States are incorporated into their legislation, and therefore SARPs, when adopted should similarly become domestic law. The fundamental flaw in this argument is that multilateral treaties are not "ICAO" treaties. They are adopted by States through diplomatic conferences. SARPs are adopted by the Council of ICAO and by virtue of Article 38 of the Chicago Convention, as discussed above, are *ipso facto* impotent of prescriptive legal legitimacy. In other words, SARPs are merely discretionary.[22]

Connected to this issue is the right of the aggrieved for information of their diseased relatives and loved ones and how they died and the exact cause for their death. The Dutch Safety Board, which is investigating Flight MH 17 states that the main objective of the international investigation is to provide next of kin and the (international) community with information regarding the cause of the crash. All available information has to be used.[23] The implicit message here is that in such an investigation, the legal maxim prohibiting *suppressio veri* or *suggestio falsi* (suppression of the truth and suggestion of falsehood) makes for penal sanction. Any person who deliberately withholds information should be held accountable under the law at pain of punishment. In this context a distinct separation has to be made between human error and wilful misconduct (which is of a higher degree than gross negligence) where, although evidence of neither can be suppressed, the former is excused on the basis of the lack of *mens rea* (the mental element) involved in the act. Therefore what is needed is not for ICAO to ascribe to itself the authority of a legislator (which it is not) but to examine the principles that would lead to convincing States to adopt legislation that could balance the absolute insistence on information as a legal requirement and mitigatory circumstances that would instil confidence of impunity in a source of information.

In the case of an accident investigation, the liability of the carrier and its servants, which is of primary concern in the instance of eliciting safety related

[21] Article 38 provides *inter alia* that any State which finds it impracticable to comply in all respects with any such international standard or procedure, or to bring its own regulations or practices into full accord with any international standard or procedure after amendment of the latter, or which deems it necessary to adopt regulations or practices differing in any particular respect from those established by an international standard, shall give immediate notification to ICAO of the differences between its own practice and that established by the international standard.

[22] The only exception is Annex 2 to the Chicago Convention on Rules of the Air.

[23] http://www.onderzoeksraad.nl/en/onderzoek/2049/investigation-crash-mh17-17-July-2014/onderzoek/1553/what-is-the-dutch-safety-board-investigating.

information, is governed by either the Warsaw Convention of 1929[24] or the later Montreal Convention of 1999[25] which replaced the Warsaw Convention. Both treaties impose a presumption of liability on the carrier and the former treaty has a limitation of liability which can be lifted in the case of wilful misconduct of the carrier. The Montreal Convention has a system of liability that is based on compensation of 113,100 Special Drawing Rights (SDRs). For proven damages in excess of 113,100 SDRs the carrier has the option of proving it had no fault in causing the Article 17 "accident" or that the Article 17 "accident" was the sole fault of some third party, in order to avoid payment of the proven damages in excess of 113,100 SDRs.

On this basis, if a staff of a carrier who is questioned in relation to an accident, involving that carrier has withheld crucial information, which if revealed before the accident would have prevented that accident, the carrier would be liable for wilful misconduct. That would make the information source naturally reluctant to come forward with knowledge of such circumstances fearing reprisals or criminality. At an information session held by ICAO just prior to the opening of the Second High Level Safety Conference, a panel wondered whether such safety information should be totally protected, giving an information source the comfort of immunity from penal sanctions, thus ensuring that all necessary evidence is in the open that would facilitate the accident investigation and the determination of the cause of the action, or whether all such protection should be removed, which may discourage information sources from revealing facts that might incriminate them. No conclusion was reached at the information session with some agreeing with the former and others in the panel agreeing with the latter proposition.

This dichotomy can be applied to the accident of Air Asia flight 8501[26] and the fact that there are two key words that drive aviation safety—standardization and harmonization. Standardization means compliance with regulations, processes and practices. Harmonization means global consistency in regulations, processes and practices. Without these words being put into practice safety in aviation would inevitably fail. The Air Asia flight operated despite authorization having been withdrawn for the flight that Sunday. The withdrawal notice would have been communicated to the persons responsible for the flight, yet it was operated. At the

[24] Convention for the Unification of Certain Rules Relating to the International Carriage by Air, Signed at Warsaw on 12 October 1929, Schedule to the United Kingdom Carriage by Air Act, 1932; 22 & 23 Geo.5, ch. 36, reprinted in Annals of air and Space Law 2005 Vol. XXX Part 1 at 325.

[25] Convention for the Unification of Certain Rules for International Carriage by Air, signed at Montreal on 28 May 1999, ICAO Doc. 9740.

[26] On Sunday 28 December of 2014 Air Asia flight QZ 8051 crashed into the Java Sea on its way from Surabaya to Singapore, killing all 162 passengers and crew on board. It was later reported that the flight did not have authorization to be operated on the route on Sundays since authorization had been withdrawn by the Indonesian authorities. Later, on 2 January 2015, the Indonesian authorities suspended Air Asia flights departing from Indonesian airport and brought in strict regulations and even stricter monitoring and supervision controls in its air transport system.

accident investigation this issue would come up and those responsible may be reluctant to reveal their knowledge of the despatch of the aircraft for fear of criminalization. The issue would revolve round the two words standardization and harmonization and it is critical that the entirety of evidence is known with regard to a particular accident.

It is submitted that in the first and final analysis, the suppression of evidence in the interest of finding the cause of the accident with a view to preventing future accidents, or the protection of persons to whom culpability can be ascribed, although worthy of consideration, cannot be used to subvert the purpose or process of justice or the punishment of those guilty of intentional wrongdoing. Anyone who is guilty of committing a wrongful act with the mental element recognized by law should be subject to the law. This is the rule of law which stands supreme over every other human endeavour or conduct. One need not feel threatened if one follows the rules of standardization and harmonization and, of course, if there is human error without negligence involved, there would be a natural exemption and protection from the penal sanctions of justice.

4.4.3 GermanWings

On 23 March 2015 *GermanWings* Flight 9525 bound from Barcelona to Dusseldorf crashed into the French Alps, all other 150 people in the aircraft. The deceased co pilot is accused of deliberately "destroying" the Airbus A320. He is reported to have crashed the aircraft, by taking control of the aircraft while the captain had left the cabin for a brief while, and locking the captain from the flight deck on his return. The captain is recorded (by the cockpit voice recorder that was subsequently retrieved) as banging forcefully on the door to get in. Normal breathing from within the cockpit is also recorded indicating that the co pilot was not incapacitated or ill.

There have been a few instances in the past where pilots are alleged to have deliberately crashed the aircraft they were piloting. One was the EgyptAir crash of Flight 990 from New York to Cairo on 31 October 1999 about 60 miles south of Nantucket, which killed all 217 aboard. The accident investigation report revealed that there were no problems with the aircraft but a relief pilot had steered the aircraft into the ocean minutes after takeoff.

Two years earlier, SilkAir Flight 185 flying from Jakarta to Singapore on 19 December 1997 crashed into the Musi River in Indonesia killing all 104 people aboard. Again, no mechanical failure was found and accident investigators concluded that the plane was probably steered to the river from the cockpit.

On 9 February 1982 a Japan Airlines jet crashed into Tokyo Bay on approach to Haneda Airport killing 24 of the 174 people on board. The crashed was attributed to the captain who was declared mentally unstable.

More recently, one theory that has been forwarded about the missing Malaysian Flight MH 370 which has still not been found since its disappearance during a flight from Kuala Lumpur to Beijing on 8 March last year, is pilot suicide.

On 29 November 2013, a Mozambique Airlines aircraft crashed in Northern Namibia, killing all 27 passengers and six crew members. Here, it was revealed that the captain had steered the aircraft to the ground after locking the co pilot out of the cockpit.

On 21 August 1994 a Royal Air Maroc flight crashed into a mountain after takeoff from Agadir, Morocco, killing all 44 on board. The investigative report revealed that the pilot crashed the plane deliberately with a view to committing suicide.

It s ironic that the aviation community decided to strengthen cockpit doors against terrorist attacks after the heinous crimes by terrorists who used aircraft as weapons of mass destruction on 11 September 2001, only to find fortified doors now facilitating criminal acts of flight crew.

In the context of the *GermanWings* disaster, the airline is reported to use a protocol that identifies a person outside the cockpit with a closed circuit TV, but a pilot could remain alone in the flight deck. Some experts in the United States have criticized *GermanWings* for keeping the co-pilot (who had only 630 flying hours under his belt) alone in the flight deck, as in the United States no pilot having less than 1000 flying hours would be allowed to take control of even a small commuter aircraft.

It has been reported that investigators have discovered much evidence from the co-pilot's home which point to the fact that he was under medical treatment for mental illness. This is indeed disturbing. The integrity of aircrew in the medical equation is entirely dependent on the question "are you fit to fly"? It is only the crew member who can answer that question. Thus, it boils down to a matter of trust between the crew member and operator.

International regulations adopted under the auspices of the ICAO require that a pilot has to have a certificate of competence issued by the State in which the aircraft he flies is registered, if he were to undertake flying an aircraft. This requirement is articulated in Article 32 of the Chicago Convention of 1944. Medical certification is an essential component in the licensing process and conditions and guidelines for the issuance of such certificates are provided in detail in ICAO documents. The overall responsibility of the pilot for the safety of his flight and that of persons therein which is legally recognized by international treaty, has necessitated the grounding of pilots for many reasons where their health did not reach the standards required, which in turn has resulted often in the concealment by pilots during their medical examinations of pre-existing medical conditions.

The pilot operates in a highly complex environment, particularly in single pilot operations. Contemporary commercial airline practice and the tenets of air law attribute to the pilot of an aircraft absolute responsibility for the safe operation of his aircraft. This responsibility can be carried out only if the pilot is not negligent or if he enjoys basic health as required by applicable regulations. The International federation of Airline Pilots Associations (IFALPA) has a code of ethics for airline pilots which stipulates *inter alia* that a pilot will not knowingly falsify any log or record nor will he condone such action by other crew members. Furthermore, the code requires the pilot to keep uppermost in his mind that the safety, comfort, and well-being of the passengers who entrust their lives to him are his first and greatest responsibility. The pilot also undertakes that he will never permit external pressures or personal desires to influence his judgment, nor will he knowingly do anything that could jeopardize flight safety and that he will remember that an act of omission can be as hazardous as a deliberate act of commission, and he will not neglect any detail that contributes to the safety of his flight, or perform any operation in a negligent or careless manner.

These ethics impliedly require a pilot to divulge to his employer and insurer any medical condition and medications taken to treat that condition. The typical pilot's disability insurance coverage is given upon the pilots assurance *inter alia* that he is not aware of any deterioration in general health, hearing, eyesight or blood pressure, and that in the event of any fraud, misstatement, concealment or failure to disclose information in response to any question, whether intentional or inadvertent, the coverage given will become void and no benefits will be payable.

There have been several instances where pilots have either falsified or concealed their medical history. One commentator records 46 instances of pilots in Northern California in the United States who did not disclose their debilitating health to the Federal Aviation Administration (FAA) that would have disqualified them from obtaining their pilots licences. The pilots in question had claimed to be medically fit to fly, yet at the same time were receiving social security payments for medical disabilities. This discovery was a result of an investigation started by the FAA. In July 2003, when, the Department of Transportation Office of Inspector General (DOT-OIG) and Social Security Administration Office of Inspector General (SSA-OIG), citing safety and security concerns, initiated a joint investigation to identify pilots misusing Social Security numbers. During the course of the investigation, social security records identified individuals who also held FAA medical certificates and who were receiving Social Security Administration (SSA) disability benefits. The DOT-OIG and SSA-OIG launched an 18-month probe termed "Operation Safe Pilot" in coordination with the U.S. Attorney's Office (USAO) to look into possible fraudulent activity. 40,000 pilots were suspected of lying or falsifying their medical history and it was discovered that 3220 pilots with current medical certificates were collecting SSA benefits, including disability benefits.

It has been recorded that a surprisingly significant number of pilots face denial of medical certification because they are taking antidepressant or serotonin blocker drugs (SSRI's) such as Prozac, which could imperil a pilot's functions. Medical certification requires that airmen be able to exercise the duties privileges of a pilot in the class applied for. In addition, numerous medical conditions will disqualify a pilot from obtaining medical certification, including, *inter alia*: diabetes mellitus, myocardial infarction, cardiac valve replacement, permanent cardiac pacemaker, personality disorders that are severe enough to have repeatedly manifested itself by overt acts, substance dependence or abuse, and epilepsy.

As long as the crew member suspects or has reason to believe that such disclosure would result in punitive measures or loss of pay (which could even lead to loss of career) he/she would not usually volunteer information even if such inaction may jeopardize the safety of the flight. Such is human nature. One approach to this question would lie in the triumvirate of regulator, operator and crew member reaching a cohesive relationship of trust that would obviate jeopardising the interests of career and more importantly the interests of safety. Attitudes should change, resulting in a culture change. The regulator must ensure that there are regulations in place that would protect both the operator and crew member. The following elements are critical to this process: shared responsibility; acknowledgment of the complexity of illness in flight crew and the fact that illness, as a risk factor can never be totally eliminated; multiple solutions for multiple problems; scientific progress; and continuous evaluation of both physical and mental health with a view to enhancing guidance.

4.4.4 Dissemination and Sharing of Safety Information

The ICAO Secretariat recommended to the Conference that ICAO develop [the] (sic) global input sharing framework that can be used for different types of information including but not limited to, exchange of operational input.[27] ICAO anchored itself, for the basis for this recommendation, on Article 54 (i) of the Chicago Convention which recognized as a mandatory function of the Council—that it request, collect, examine and publish information relating to the advancement of air navigation and the operation of international air services, including information about the costs of operation and particulars of subsidies paid to airlines from public funds. In its submission to the Conference ICAO linked this function to a permissive function appearing in Article 55 (c) which provides that the ICAO Council may conduct research into all aspects of air transport and air navigation which are of international importance, communicate the results of its research to the contracting States, and facilitate the exchange of information between contracting States on air transport and air navigation matters, together with Article 55 (i) which provides that ICAO may investigate, at the request of any contracting State, any situation which may appear to present avoidable obstacles to the development of international air navigation; and, after such investigation, issue such reports as may appear to it desirable.[28]

The three provisions cited do not mesh, as the first—the mandatory function of the Council in requesting, collecting and publishing information, is a function where ICAO operates as a "mail box". The function in Article 55 (c) gives ICAO the discretion to conduct research into all aspects of air transport and air navigation and communicate the results of its research—a function totally different from the first cited function. Quite curiously, the third function cited is even more disparate—that ICAO investigate, at the request of any contracting State, any situation

[27] Evolving Safety Analysis to Support Global Aviation Safety Strategies, *HLSC/15-WP/5*, 29/9/14.

[28] Id. at 1–2.

which may appear to present avoidable obstacles to the development of international air navigation; and, after such investigation, issue such reports as may appear to it desirable. There are three functions here—ICAO as a collector of information and distributor of such; ICAO as a researcher and ICAO as an investigator.

Apart from this startling disconnect in the basis for ICAO's request to be a focus of global input sharing framework, the fundamental flaw in the citation of Article 54 (i) is that the main aim of ICAO, as prescribed in Article 44 of the Chicago Convention is to develop the principles and techniques of air navigation. The "advancement of air navigation and the operation of international air services" is intrinsically linked to the development of the principles and techniques of air navigation. In its enthusiasm to function as a mail box the ICAO Secretariat seems to have entirely missed the point of ICAO's sole aim and objective under the Chicago Convention.

Even more intriguing is that in a separate submission,[29] ICAO advised the Conference that a centralized global information capability, combining and replicating information provided by Contracting States and relevant agencies on risks to civil aviation arising from conflict zones supports the risk assessment process of parties concerned. A caveat to this submission was that ICAO shall not be liable for any direct, indirect, punitive, incidental, special or consequential damages (including, but not limited to, damages for loss of business profits, business interruption, loss of programmes or information) that result from the use or inability to use such site, in particular for, but not limited to, errors or omissions in the contents of the website or the consequences of its use, nor for inaccurate transmission or misdirection, even if ICAO has been advised of the possibility of such damage.[30] One could inquire as to how any responsible State or operator would rely on information provided which has no warranty of authenticity, responsibility or accountability.

The Russian delegation had a practical approach to the issue which they orally submitted during the floor discussions. The delegation recognized that it was appropriate to set up a globalized or a centralized information sharing facility as regards the use of airspace and potential threats to flights. However, in order to achieve this objective the Russian delegation felt that a legal assessment of all relevant aspects regulating the activities of such a system was a compelling need for a globalized system to be finalized. Of particular concern was the suggestion to include a provision disclaiming responsibility. The Russian Federation felt strongly that the issue of a global information capability should be addressed by an ICAO Assembly and approved thereby. The delegation recommended that ICAO carry out additional work on the development and testing of a prototype information sharing system based on the provisions of the concept of information sharing and called on States to share information on risks affecting conflict zones and adjacent zone using

[29] Sharing Information on Risks to Civil Aviation Arising from Conflict Zones, *HLSC/15-WP/13*, 2/1/15.

[30] *Id.* Appendix C.

NOTAMs circulars and AIPs or supplementing the circulars containing aeronautical information.

In its discussion of this issue, the Conference noted progress made on the development of an initial prototype centralized capability to support the exchange of available information in support of such risk assessments and strongly supported the development of such a capability and indicated that a simple web-based system should be developed and implemented without delay.[31]

4.4.5 Legal Issues

It is not entirely clear from the discussions of the Conference and working papers submitted as to the nature or thrust the centralized capability to support the exchange of available information on conflict zones would be. One international Organization suggested that there could be SARPs incorporated in Annex 19 on Safety Management which ICAO can insist that States incorporate in their laws. This has already been discussed in this paper. It is worthy of note that Latvia (on behalf of member States of the European Union, the European Civil Aviation Conference and EUROCONTROL) submitted *inter alia* that a study, carried out by the ICAO Filing of Differences Task Force (FDTF), identified a widening gap between the adoption of Standards and Recommended Practices (SARPs) by ICAO and their implementation by States. It was claimed that this widening gap was cause for serious concern when the aviation community at large should be improving the overall implementation of SARPs and thus improving overall safety management. The EU working paper went on to say that the issue of interpretation and implementation has been identified as a continuing problem in this context, concluding that some form of quality control was required to ensure that the apparent lack of clarity does not have a negative impact on the proper implementation of SARPs, especially when consideration in the overall safety management context is being given to the introduction of performance-based regulations. It was observed that this was in part because there is no effective feedback process.[32]

Another confusing anomaly is what ICAO's exact role is in the establishment and maintenance of this centralized capability. Ensuring safety is essentially a State

---

[31] HLSC/15-WP/102 *Revised* at 3. It must be noted that the Task Force appointed by ICAO to address the issue of a centralized capability to support the exchange of available information has advised that ICAO act as host of the information and that, in order to do so, and depending on the structure and complexity of the proposed system, ICAO would need to engage the services of additional staff, including senior personnel having signing authority, and comply with all staff rules regarding leave, training, etc. The Secretariat estimates that these costs, along with those required to supply essential office infrastructure (space, computers, etc.) and continuous IT support service, will require substantial contribution of funds from the member States of ICAO up to US $2.5 million. See *HLSC/15-WP 33* at p. 4.

[32] *Towards a Global Implementation of Safety Management*, HLSC/15-WP/35, 15/12/14 at 4.

responsibility issue and ensuring safety cannot be thrust upon any other entity which could only collect information and maintain such information in a repository with a disclaimer of responsibility in tow. ICAO Assembly Resolution A37-1 (Principles for a code of conduct on the sharing and use of safety information) provides that ensuring the safety of international civil aviation is the responsibility of Member States both collectively and individually. An earlier Resolution (still in force) A29-13 (Improvement of safety oversight) reaffirmed that one of the tenets of the Chicago Convention is that it is an individual State's responsibility for safety oversight. Nine years later, when ICAO established its Aviation Safety Oversight Audit Programme through Resolution A32-11 (Establishment of an ICAO Safety Oversight audit Programme), the Assembly recognized that the objectives of the ICAO safety oversight programme were to seek to ensure that Contracting States are adequately discharging their responsibility for safety oversight over aircraft operations, the licensing and training of personnel, and aircraft airworthiness. The same resolution recalled that ultimate responsibility for safety oversight rests with Contracting States, which are required to continuously review their respective safety oversight capabilities.

At the latest Assembly held in 2013 (38th Session) this overarching principle of State responsibility was reiterated in Resolution A38-5 (Regional cooperation and assistance to resolve safety deficiencies, establishing priorities and setting measurable targets)—that ensuring the safety of international civil aviation is the responsibility of Member States both collectively and individually. However, although States have overall responsibility for safety in aviation, they are not alone. ICAO and other key players are required to collaborate when States implement ICAO's Global Aviation Safety Plan. The 37th Session of the assembly, held in 2010, adopted Resolution A37-5 (The Universal Safety Oversight Audit Programme (USOAP) continuous monitoring approach) which, in a preambular clause recognizes that safety oversight, and the safety of international civil aviation in general, is the responsibility of Contracting States, both collectively and individually, but it also depends on the active collaboration of ICAO, Contracting States, industry and all other stakeholders in the implementation of the Global Aviation Safety Plan. The bottom line in responsibility is explicitly stated in Resolution A32-19 (Charter on the Rights and Obligations of States Relating to GNSS Services) which recognizes that every State preserves its authority and responsibility to control operations of aircraft and to enforce safety and other regulations within its sovereign airspace.

4.4.6 Consumer Protection

Be it the protection of safety information or the sharing thereof, ICAO and the rest of the aviation community must take cognizance of the fact that ultimately such should be in the best interest of the consumer of the air transport product and his/her interests including the interests of those closest to the victim of an accident. It is incontrovertible that at the centre of all this is the consumer of the air transport product.

ICAO has not helped at all on the subject of consumer protection except for an ambivalent and meaningless nod to States at its 5th Air Transport Conference 12 years ago:

> As a premise in addressing consumer interests issues, States need to carefully examine what elements of consumer interests in service quality have adequately been dealt with by the current commercial practices of airlines (and service providers, if applicable) and what elements need to be handled by the regulatory and/or voluntary commitment approaches. States need to strike the right balance between voluntary commitments and regulatory measures, whenever the government intervention is considered necessary to improve service quality. . .[33]

The United Nations on the other hand has adopted guidelines on consumer protection which apply in general to the consumer. By Resolution A/RES/39/248 of 16 April 1985 (Consumer Protection) to which is Annexed the guidelines the United Nations recognized that: in taking into account the interests and needs of consumers in all countries, particularly those in developing countries; recognizing that consumers often face imbalances in economic terms, educational levels, and bargaining power; and bearing in mind that consumers should have the right of access to non-hazardous products, as well as the right to promote just, equitable and sustainable economic and social development, these guidelines for consumer protection have the following objectives: to assist countries in achieving or maintaining adequate protection for their population as consumers; to facilitate production and distribution patterns responsive to the needs and desires of consumers; to encourage high levels of ethical conduct for those engaged in the production and distribution of goods and services to consumers; to assist countries in curbing abusive business practices by all enterprises at the national and international levels which adversely affect consumers; to facilitate the development of independent consumer groups; to further international co-operation in the field of consumer protection; to encourage the development of market conditions which provide consumers with greater choice at lower prices.

The Resolution also states that Governments should develop, strengthen or maintain a strong consumer protection policy, taking into account the guidelines set out below. In so doing, each Government must set its own priorities for the protection of consumers in accordance with the economic and social circumstances of the country, and the needs of its population, and bearing in mind the costs and benefits of proposed measures.

The Resolution furthermore provides that the legitimate needs which the guidelines are intended to meet are the following: the protection of consumers from hazards to their health and safety; the promotion and protection of the economic interests of consumers; Access of consumers to adequate information to enable

[33] Report of the Worldwide Air Transport Conference, Challenges and Opportunities of Liberalization, (Montreal, 24–28 March 2003) *ICAO Doc 9819; ATConf/5 2003* at 42.

them to make informed choices according to individual wishes and needs; consumer education; availability of effective consumer redress; freedom to form consumer and other relevant groups or organizations and the opportunity of such organizations to present their views in decision-making processes affecting them.

In the Guidelines is the fundamental principle that Governments should adopt or encourage the adoption of appropriate measures, including legal systems, safety regulations, national or international standards, voluntary standards and the maintenance of safety records to ensure that products are safe for either intended or normally foreseeable use. Also in the Guidelines is a statement to the effect that Government policies should seek to enable consumers to obtain optimum benefit from their economic resources. They should also seek to achieve the goals of satisfactory production and performance standards, adequate distribution methods, fair business practices, informative marketing and effective protection against practices which could adversely affect the economic interests of consumers and the exercise of choice in the market-place.

The Guidelines go on to say that Governments should intensify their efforts to prevent practices which are damaging to the economic interests of consumers through ensuring that manufacturers, distributors and others involved in the provision of goods and services adhere to established laws and mandatory standards. Consumer organizations should be encouraged to monitor adverse practices, such as the adulteration of foods, false or misleading claims in marketing and service frauds. Governments are also called upon to develop, strengthen or maintain, as the case may be, measures relating to the control of restrictive and other abusive business practices which may be harmful to consumers, including means for the enforcement of such measures. In this connection, Governments should be guided by their commitment to the Set of Multilaterally Agreed Equitable Principles and Rules for the Control of Restrictive Business Practices adopted by the General Assembly in resolution 35/63 of 5 December 1980.

In addition, Governments are called upon to adopt or maintain policies that make clear the responsibility of the producer to ensure that goods meet reasonable demands of durability, utility and reliability, and are suited to the purpose for which they are intended, and that the seller should see that these requirements are met. Similar policies should apply to the provision of services. Governments should also encourage fair and effective competition in order to provide consumers with the greatest range of choice among products and services at the lowest cost.

These principles should be entrenched and enforced in a regime dedicated to the consumer of air the transport product. Perhaps the CANSO *Just Culture*[34] could be a starting point.

The Second ICAO High Level Conference succeeded in serving as a global forum of discussion for States on key issues pertaining to safety. 850 participants from States (at the DGCA level) and international organizations voiced consensual views on many of the issues discussed, including the importance of protecting

[34] *Supra* note 18 (this chapter).

safety information and providing information on risk assessment pertaining to conflict zones, collected from States. The Report of the Conference will go before the ICAO Council for acceptance and determination of tasks would follow from Council directives.

This chapter has discussed numerous anomalies that arose from the discussions and recommendations. Perhaps the greatest anomaly is one which has been, and continues to be perpetuated—the rights of the consumer. In this context two questions arise. Who should be protected in terms of the dissemination of accident information—the information source or the injured passenger and the dependents of deceased passenger? Should information on wilful misconduct be suppressed, depriving the claimant of proper compensation? Should the prevention of accidents be reactive rather than proactive? With regard to information regarding conflict zones, should not the passenger have the right to expect from the State (and not an international organization) the protection from being shot down?

The ICAO Council has a practice of endorsing a report submitted by an ICAO conference and requesting the Secretary General to pursue recommendations therein and report back to the Council. What the Council should be doing is, giving detailed consideration to the report and, in addition to what is reported, think independently as to what should be done to address key issues raised in such report. In this instance, the conference made several recommendations, mostly seeking harmonization in protecting and sharing information.[35] It is submitted that issues raised in this article could be of critical interest to the Council in view of the chronology of breaches in safety, ranging from pilot malfeasance which have already been chronicled herein, to technological, political and economic issues that would primarily affect the interests of the consumer of air transport—a subject that seems to allude those having responsibility to ensure such interests.

References

Abeyratne R (2014) Flight MH 17: the legal and regulatory fallout. Air Space Law 39(6):329–342

Abeyratne R (2015) Flight MH 370 and global flight tracking – the ICAO role in risk management. Zeitschrift fur Luft- und Weltraumrecht (German J Air Space Law) 64:2–51

[35] Second High Level Safety Conference 2015, Report, Doc 10046, HLSC 2015 at 3-1.

Chapter 5
Competition and Innovation in Air Transport

5.1 Introduction

5.1.1 Competition

Air transport is a different animal from other aspects of trade, in that it is circumscribed by the need for airlines to obtain permission of the grantor State to obtain market access. In this context, how does one approach market access in the coming decades? At the most fundamental level, the advantages of free trade as would apply to air transport would be that it would encourage States to trade freely with their trading partners which would in turn help in the growth of the global economy; it would give the consumer a better choice of products and competition generated by free trade would bring down the price of the product. Arguments against free trade in air transport would be that globalization and liberalization will take jobs away from a State; the limit of imports would keep money in the State; free trade could be a threat to national security and a State could develop dependence on the expertise of other more advanced States. Free trade increases national wealth and promotes foreign investment, both of which are absent in the present structure of market access in many States.

The main consideration, leading up to efforts by the international aviation community to achieve a deregulated global airline industry, is involved with the question as to whether free market principles can be applied globally to air transport. What needs to be considered is whether we are ready to accept the throwbacks as consequences of free market competition in air transport, particularly in losing national prestige projected by flag carriers. One of the corollaries to industry deregulation is the introduction of free market competition when companies switch from operative performance to competitive performance. Competition therefore emphasizes the need to focus on a company's performance in relation to its competitors. This principle can be readily apply to various industries that have already been deregulated, such as the motor vehicle industry, chemical industry and

R. Abeyratne, *Competition and Investment in Air Transport*,
DOI 10.1007/978-3-319-24372-6_5

information technology industry. The operative question is "are these good analogies for application to the air transport industry?" Whatever be the answer to this question, if the deregulated domestic air transport industry of the United States were to be considered an analogy, one could say that a deregulated system in the United States, introduced in 1978, has led to a more efficient airline system in the country. Whatever be the case, access to facilities in a competitive market is essential toward attaining fluidity of market forces. In the air transport industry, this can be translated to mean that if free markets do not exist in the supply of complementary facilities, there will be no positive impact of liberalization. The complementary services in the supply of air transport are airport access, computer reservation systems and airport and air regulation services.

The International Chamber of Commerce (ICC), in a policy statement has expressed the view that the efficiency of air transport would be enhanced by creating more open markets and more flexibility with regard to foreign ownership. Given air transport's capability to facilitate economic activity, its liberalization would enable the sectors that make use of it to become more efficient. ICC was in favor of a freer exchange of air services throughout the world and is convinced that it is time to move beyond the existing bilateral system, toward a genuine multilateral liberalization of air transport Of course, liberalization would give way to competition, which in turn would impel airlines to pool their resources in order that they maximize on such assets as code sharing and airport slots. However, alliances do not necessarily mean lack of competition between partners. Airlines within alliances have to do their utmost to gain market access and keep their businesses alive. In order to do this both private enterprises and the States in which these enterprises are entrenched have to be equally competitive.

Any agreement to bring in an aspect of trade within a liberalized framework is generally a pro-active measure, which brings to bear the willingness and ability of the governments to face trading issues squarely in the eye. However, any agreement for trading benefits would be ineffective without the element of competition, both between enterprises and between States. The essential requisite for success in trading relations is competition, which in turn leads to national prosperity. A free trade agreement is merely the catalyst in the process.

The shift in focus that confronts competition in commercial aviation largely lies in the compelling need for measures that would enable airlines to maximize on potential available in the market, thereby preventing the industry from falling apart. Since the beginning of regulated civil aviation in 1944 with the signing of the Chicago Convention and until recently, competition was rigidly regulated and often based on predetermined capacity as a determinant of a carrier's entitlement to enter a market, which stultified the global air transportation system. The transition from a somewhat smoothly running but cumbersome aviation industry was not easy, as the air transport industry was comfortable and languid with established legacy carriers dominating a vastly untapped market. However, the then air transport system had its advantages. There may not have been large aircraft and more connections than there are today, and notwithstanding the absence of the internet where airline tickets and hotel reservations could now be obtained at the flick of a hyperlink on the home PC, in the past a passenger almost anywhere in the world could purchase a ticket to fly seamlessly to almost any part of the world, through a complex but reasonably efficient set of working relationships between hundreds of individual air carriers. Transaction costs were low, where a single call to a travel agent at the corner of the street could finalize a trans-continental flight. This was mostly possible because individual airlines themselves ensured the provision of all their services, infrastructure and procedures to connect passengers and freight both within their own networks and those of connecting carriers. At the airport, the passenger did not encounter heat sensitive monitors that sought to establish that he was the carrier of a communicable disease nor were there physical checks of person and baggage during check in. Industry standards and facilitation measures were in place through the International Civil Aviation Organization (ICAO) and the International Air Transport Association (IATA). Procedures and clearance at the airport was hassle free and not subject to color coded security alerts.

Today, the story is somewhat different. Things have become more sophisticated. It was inevitable that they had to, as market conditions changed and the demand for air transportation grew (and keeps growing) at roughly double the rate of the growth in the general economy. The exponential infusion of capacity to meet this demand requires costly systems and infrastructure to serve the consumer. An inevitable corollary was that air transport became more expensive. However, if only this were the issue, things would have been easily manageable. The prolific use of air travel by the public made air transport an easy target for terrorism and the free movement of disease causing vectors.[1] The threat of terror brought its own problems requiring more expensive aviation insurance and the necessity to cancel flights every now and then, while the proliferation of air travel brought in environmental concerns of aircraft noise and engine emissions, problems of airport congestion and slot allocation. Worst still, competition between carriers to offer capacity made some of them expand with the only objective of providing services at any cost. Critical services required for aviation safety, such as efficient ground handling and precise engineering were outsourced, with no guarantee of one hundred percent safety of a flight.[2] The crisis deepened.

The above notwithstanding, neither a single nation nor the global aviation community ever deregulated safety and security. Responsibility still continues to devolve upon governments to provide additional capacity, find the money to fund safety and security inspectors and ensure that their carriers operate air services with full insurance coverage, however expensive, so that the world economy would not run aground for lack of international air services. Fortunately, all 191 States who are signatories to the Chicago Convention, can rely on ICAO to find regulatory solutions that would keep the crisis from becoming unmanageable. In the recent past, ICAO has regulated on safety and security, established and conducted security and safety audits on Contracting States, adopted much needed principles of

[1] Abeyratne (2002a), pp. 53–80.

[2] See Abeyratne (1999a), Chapter 7, Outsourcing and the Virtual Airline, pp. 115–126.

guidance on facilitation, assisted in preventing the spread of disease by air carriage and even developed a global aviation insurance scheme in case things go wrong in the insurance industry the way they did immediately after 11 September 2001.[3]

The preeminent regulatory challenge confronting air transport is to update policies, guidelines and other regulatory instruments to address changes in the aviation environment. Competition, when coupled with liberalization of air services which is taking place on an international scale, inevitably calls for a more open and free approach. However, it is not prudent to consider air transport services as being just another normal economic activity. The overarching objective of ICAO, as contained in Article 44 of the Chicago Convention is for ICAO to foster the planning and development of international air transport so as to "meet the needs of the peoples for safe, regular, efficient and economical air transport". This is a most fundamental challenge which not only draws the inference that air transport is a public utility, but also issues a challenge to ICAO, its Contracting States and their carriers to ensure the provision of a safe service satisfying fixed standards of continuity, regularity, capacity and pricing.

In order to face the exponential growth of the air transport industry, it is inevitable that competition and liberalization should be given serious consideration as a current and future trend in the aviation field. What is needed foremost, in order to improve international cooperation toward achieving a well meshed and overall competitive policy, is to consider the various possible options available. One of the options to promote competition and facilitate trade in air transport lies indisputably in combating and eliminating anti competitive practices.[4] State responsibility toward achieving this goal is a key factor. One way of ensuring collective State action in this regard might be for States to enter into understandings or agreements toward combating restrictive trade practices, either bilaterally or plurilaterally. Along with a plurilateral framework of competitive policy, there also should be a concomitant bilateral structure of individual agreement between States to stringently monitor anti competitive conduct. This can only be achieved with a robust and effective international legislative structure. The important role played by public international law in this regard cannot be denied.

As for liberalization of air transport, there has so far been no indication that any State favours total liberalization calculated to open out its domestic market. Strategic alliances between airlines, be they through mergers or other arrangements, will be viewed with caution and objectivity by individual airlines and States so as to preclude the total overrunning of local interests. It is this consideration that would make liberalized ownership and control criteria less attractive to local entrepreneurs who would not encourage foreign ownership to encroach local control airlines have of their own markets.

---

[3] Abeyratne (2005), pp. 117–129. Also, Abeyratne (2002b), pp. 406–420 and also Abeyratne (2002c), pp. 521–533.

[4] Abeyratne (2001), pp. 622–623. Abeyratne (1984), p. 121.

The operative question—as to whether the global community proceeds to consider the operation of air transport services as a trading activity along the lines of other conventional trading activities or whether it should continue to consider it as a public utility which has broader connotations to States politically and therefore should remain regulated—remains one of fundamental importance to the future of international civil aviation. Therefore, liberalization of air transport cannot be dismissed as not being a viable prospect for the future, particularly in trading terms, and the players concerned must necessarily view air transport in its entirety, as a service composed of critical factors that are inherent in safe and efficient air transport.

The two integral areas that will carry the sustainability of air carriers and assurance of air services in the years to come will be regulatory control and economic strategy.[5] From an economic perspective, it is inevitable that competition will be between airline alliances rather that individual carriers. Markets will be unstable, and, in the case of individual airlines, only those who go "back to basics" to offer the consumer a service as "value for money" will survive. Ethical and moral consideration of economics, in terms of strategic airline management that provides for quality customer service, will play a major role in airline sustenance and will be the bottom line for the years to come.

From a trading perspective, both States and carriers must share equal responsibility to ensure continuity of air transport services. The uniqueness of the operation of air transport services as a trading practice lies in the symbiosis required for its sustenance between States and carriers. This peculiar relationship requires that a certain responsibility devolves upon States to ensure the prosperity of its air transport industry and to prevent the industry from collapsing. Although air transport may be heavily privatized in some instances, particularly in the developed world, it does not take away the overall regulatory supervisory role of the State and its obligation to support its carriers.

From a regulatory perspective, the challenges faced in the economic field, both in the short term and from a long term perspective, are to update and promote ICAO policies and guidelines to meet the demands of a changing environment and to seek a balance between promoting economic growth in the industry, advancing civil aviation, and strengthening security measures and facilitation.[6] In this respect, ICAO should indeed go a step further as it should, in asserting itself as an Organization that fosters the development of air transport by playing a lead role in prescribing an appropriate philosophy and principle of liberalization that would promote more competition among air carriers which should be able to offer connectivity untrammelled by overt State control and protectionism leading. In order to address these challenges, all players involved need to seek a harmonious relationship in the industry between a liberalized economic regulatory framework and proper safety, security, social and labour standards.

[5] Abeyratne (2003), pp. 3–37.

[6] Abeyratne (2002d), pp. 83–115.

Economic activity in air transport, particularly in the movement of aircraft between States territories, requires to be viewed in the context of sustainable development, where environmental protection would play a key role. Although air transport does not pose catastrophic environmental consequences on a short term basis, it has now become opportune to address trade in air transport and its effect on global environmental welfare as a composite whole, rather than within a fragmented framework. The economic aspects of environmental protection, particularly in the areas of noise charges and emissions trading as a market based option, is an inevitable challenge, particularly when it concerns global consensus on criteria for levying such charges and their quantification.[7]

The enemy of competition in air transport is complacency. Management competence, and a substantial resource base are essential for sustenance in a highly competitive air transport industry. Credibility and the strategic creativity of a brand should support the competitive efforts of an airline. A clear understanding of strengths and weaknesses are also essential prerequisites in a competitive environment. One of the most important factors in success at competition is the clear understanding of rules. An example of the last factor is seen in the response of *Etihad* in defence of its operations to the United States which was necessitated by the claim of unfair competition laid against the airline by three United States carriers:

> Etihad does not dispute the fact that over the past 10 years Etihad has increased its market share between the United States and the ISC; we embrace it. The Big 3 Carriers claim that Etihad has "taken passengers," giving the impression that passengers are some type of chattel to which legacy carriers are entitled. We think differently. We consider our passengers to be our guests, and we work very hard to earn their patronage. Neither Etihad, the Big 3 Carriers, nor any other airline are "entitled" to any passenger, and any airline that argues to the contrary perhaps exposes a more obvious reason for their loss of passenger market share: a sense of entitlement and a complete disconnect between what they offer and what customers want.[8]

5.1.2 Principles of Competition

Low cost and product differentiation are two key drivers of competitive strategy that ensure a competitive edge and position among others in the industry. In a liberalized market, such as in the open skies regime in air transport, these drivers show even more relevance where disruption of an existing product by innovation could capture an existing market as well as create potential markets that are waiting to be tapped. They would also effectively preclude new entrants from dislodging innovative enterprises. The successful carriers, particularly in the Gulf, have

[7] Abeyratne (1999b), p. 232. Also Abeyratne (1999c), p. 17. Gander and Helme (1999), pp. 12–14, 28–29. Elgar (2000). Elgar (2001).

[8] Infra, Chap. 10, note 8 at 56.

identified the real buyer *i.e.* one who looks for ready connectivity or availability of services; quality service and value for money. These carriers have also identified their customers' purchasing criteria as well as the costs involved in product differentiation. Another compelling criterion of customer preference is brand and customer expectation, which in turn are dependent on technological superiority. Branding is a promise that the seller gives the buyer and which the latter expects.

It is forecast that the world economy will remain moderately stable and healthy in the near future, despite a slowdown in economic growth. In the short term, inflation may hold steady and inflation rates will probably decrease gradually. The continuing upward trend in fuel prices is likely to increase airline fixed costs and aviation will increasingly be defined in trade terms. Aviation will also be a strong candidate for trade liberalization with a firm focus on services. A compelling factor in this overall picture will be increasing pressure on governments to facilitate transnational ownership of airlines. The other key issue will be aviation and the environment in the global scenario of air transport.

All the above indicators incontrovertibly point to one central driver of future air transport competition. The issue of competition will ensure the increasing influence of global alliances and partnerships between carriers as a key element in industry strategic development where more groups of airlines will provide direction and focus. Airline management, geared towards competition, will be called upon to improve coordination, and provide integration and stability to the air transport industry, resulting in the inevitable corollary of cost reduction.

Competition in global air transport presents unique strategic issues. Global industries, such as multinational air carrier alliances, are characterized by the presence of competitors operating worldwide from home bases in different countries. The governments concerned may have deep rooted interests and objectives relating to employment and balance of payments including other factors that may not strictly be economic but nonetheless of critical importance to competition among airlines. Therefore, airlines will be increasingly called upon to conduct stringent competitor analysis through examination of the relationships between individual air carriers and their governments. The home country's industrial policy must be well comprehended, particularly in terms of political considerations which may be related to such issues as purchases of aircraft and the exchange of market rights.

A global industry, such as air transport, is one in which commercial entities offering their services view competition as global and build strategies accordingly. Therefore, it follows that competition involves a coordinated worldwide pattern of market positions, facilities and investments. Factors to be taken into account are overlap between competitors, geographic location of carriers and defensive investments in particular markets and locations so as not to let competitors gain advantages that can be factored into their overall global posture.

Competition in the air transport industry is a complex process, as there is no consensus among airline economists as to the exact nature of the industry. The demand for air services, particularly in the context of the airline passenger, is a contrived demand emerging from other demands based on activities such as business and leisure. This calls for a certain segmentation in travel where, in business travel, the passenger does not usually pay for the travel himself, whereas in leisure travel it comes out of his own pocket. Therefore, the leisure market calls for a different kind of competition, primarily based on the fare, whereas in business travel, although the fare is important, other considerations, such as facilities on board, may also play a considerable role in competition.

Those supporting the retention of regulation argue that the very nature of air transport, being either naturally monopolistic or interdependently oligopolistic, calls for regulation in order that fares are not arbitrarily raised and remain competitive. Another theory in support of regulation is that some form of control should be exercised over mushroom airlines that may sprout up to exploit a liberalized market, thus disturbing the existing balance of an integrated network. Of course, each route is a separate market in itself and would require separate consideration. Although principles of economics of scale may apply generally to airline competition, where a fact such as larger aircraft being more efficient than smaller aircraft would apply on a general basis, individual assumptions for different markets have caused the two major aircraft manufacturers, Boeing and Airbus Industrie, to concentrate on manufacturing aircraft with strengths in speed and capacity respectively.

The main consideration, leading up to efforts by the international aviation community to achieve a deregulated global airline industry, is involved with the question as to whether free market principles can be applied globally to air transport. What needs to be considered is whether we are ready to accept the throwbacks as consequences of free market competition in air transport, particularly in losing national prestige projected by flag carriers. One of the corollaries to industry deregulation is the introduction of free market competition when companies switch from operative performance to competitive performance. Competition therefore emphasizes the need to focus on a company's performance in relation to its competitors. This principle can be readily apply to various industries that have already been deregulated, such as the motor vehicle industry, chemical industry and information technology industry. The operative question is re these good analogies for application to the air transport industry? Whatever be the answer to this question, if the deregulated domestic air transport industry of the United States were to be considered an analogy, one could say that a deregulated system in the United States, introduced in 1978, has led to a more efficient airline system in the country. Whatever be the case, access to facilities in a competitive market is essential toward attaining fluidity of market forces. In the air transport industry, this can be translated to mean that if free markets do not exist in the supply of complementary facilities, there will be no positive impact of liberalization. The complementary services in the supply of air transport are airport access, computer reservation systems and airport and air regulation services.

The regulation of competition within the European Community is governed by the EC Treaty.[9] Two provisions in particular, Articles 81 and 82, contain principles

[9] The EC Treaty, also called the Treaty of Rome, was concluded in 1957 to forge an even closer union among the people of Europe. See Goh (1997), p. 15.

which outlaw anti competitive conduct. While the former prohibits the prevention, restriction or distortion of competition, the latter makes itself applicable against abuse by one or more undertakings of a dominant position within the market. While the former essentially contains provisions for agreements, decisions or practices with anti competitive effects, the latter concerns itself with abuses of a dominant marketing position. The aim of these two provisions in particular are to preclude distortion of competition within the Common Market by supplementing the basic principles enshrined in Articles 81 and 82 with substance. The goals of the Treaty in general and Articles 85 and 86 in particular are to promote the free movement of services, goods, persons and capital whilst effectively obviating barriers to trade within the community. Both these provisions relate generally to all sectors of transport unless explicitly excluded by the Treaty provisions.[10]

Article 81 prohibits as incompatible such agreements as directly or indirectly fix purchase or selling prices or any other trading conditions; limit or control production, markets, technical development or investment; share markets or sources of supply; apply dissimilar conditions to equivalent transactions with other trading parties, thereby placing them at a competitive disadvantage; and make the conclusion of contracts subject to acceptance by other parties of supplementary obligations which, by their nature or according to commercial usage, have no connection with the subject of those contracts. These conditions are imposed on agreements between undertakings, which are defined as independent entities performing some economic or commercial activity.

Article 82 provides that any abuse by one or more undertakings of a dominant position within the Common Market or in a substantial part of it shall be prohibited as incompatible with the Common Market insofar as it may affect trade between member States. The Article prohibits: direct or indirect imposition of unfair purchase or selling prices or unfair trading conditions; limitation of production, markets or technical development to the prejudice of consumers; application of dissimilar conditions to equivalent transactions with other trading parties, thereby placing them at a competitive disadvantage; and conclusion of contracts subject to acceptance by the other parties of supplementary obligations which, by their nature or according to commercial usage, have no connection with the subject of such contracts.

In implementing these two provisions, air carriers have to exercise caution in not assuming that purely in view of a bloc exemption on air transport in the Treaty that may pertain to a particular issue, a related practice would be exempt from the prohibitions contained in Articles 81 and 82. In the air transport section of the Treaty, it is abundantly clear that block exemptions may apply only if abuse of dominant position is not evident in a given transaction.[11] Articles 81 and 82 are independent and complementary provisions and any exemption under Article

[10] Case 167/73 *Commission* v. *French Republic* [1974] ECR359 at 370.

[11] See Adkins (1994), p. 81.

81 will not necessarily render the provisions of Article 82 nugatory.[12] Dominant position was defined in the 1979 decision of *Hoffman-La Roche* v. *Commission*[13] as a position of economic strength enjoyed by an undertaking which enables it to prevent effective competition being maintained on the relevant market by affording it the power to behave to an appreciable extent independently of its competitors, its customers and ultimately of its consumers. Such a position may necessarily preclude some competition except in monopoly or quasi monopoly situations. There is every indication, from existing jurisprudence and EC practice, that an assessment on an abuse of dominant position would not be predicated upon one factor alone or single characteristic but would rather be anchored on numerous factors such as market structure, barriers to entry and conduct of the business enterprise concerned.

In the United States, the term antitrust laws encompasses federal and States legislation (statutes) which regulate competition with a view to wiping out unfair trade practices and preserving competition among sellers and buyers. Needless to say, antitrust laws apply equally to international air services, and are calculated to preclude both conduct and structural changes in business enterprises. A typical example of conduct coming under antitrust laws in the United States is a merger between competitors which would unduly limit competition. These laws are also meant to prevent producers or purchasers of goods from exercising a monopoly in imposing prices which significantly deviate from expected free market competition norms.

Antitrust legislation in the United States goes back to 1890 with the enactment of the *Sherman Act* which makes it criminally illegal for any contract, combination or conspiracy to be formed in restraint of trade. This all encompassing provision prohibits price fixing, anti discounting agreements, divisions of markets by pooling agreements and capacity agreements and exchanges of information that can be considered as competitively sensitive. The Act also prohibits monopolies and conspiracy to monopolize in Sect. 5.2.

In 1914, the United States Congress legislated the *Clayton Act*, primarily to supplement the *Sherman Act*. The *Clayton Act* outlaws certain types of exclusive dealing and contrived sales and prescribes standards for determining the legality of mergers and acquisitions. Both the Acts admit of compensation to persons injured in their trade or business up to three times the amount of their loss plus attorney fees. Courts have also permitted consumer class actions an antitrust activity, allowing for significant recovery of damages.

In its role as the sole international regulatory body in the field of air transport, the International Civil Aviation Organization has issued clear policy and guidance material on the avoidance or reduction of conflicts over the application of competition laws to international air transport. ICAO has issued these guidelines to address the conflicts that may arise between States which adopt policies, practices and laws relating to the promotion of competition and restraint of unfair

---

[12] Case T-51/89, *Tetra Pak Rausing SA*. v. Commission [1990] II E.C.R. 309, [1991] 4 C.M.L.R. 334 para 31.

[13] Case 85/76 *Hoffman-La Roche & Co. A.G.* v. *Commission* [1979] E.C.R. 461.

competition within their territories. ICAO urges States to ensure that their competition laws, policies and practices, and any application thereof to international air transport are compatible with their obligations under relevant international agreements.[14] Within this guideline, there is a strong recommendation for close consultation between States and all interested parties in order that uniformity in practice be achieved across borders to the maximum extent possible. Accordingly, when a State is adopting laws pertaining to competition, it is expected to give full consideration to views expressed by any other State or States whose interests in international air transport may be affected. States are urged to have full regard to principles of international comity, moderation and restraint. The Guidelines also provide direction on dispute resolution and problem solving.

The regulation of air transport services lies within the purview of ICAO member States which maintains in its Legal Bureau a register of all bilateral air transport agreements. The bilateral air transport agreement usually includes a reciprocal agreement between States for their carriers to have fair and equal opportunity in operating air services between their territories without unduly affecting the air services operated by each other. Under a bilateral agreement, capacity offered by carriers must bear close relationship to the needs of the people using air transport.[15] These regulatory provisions have so far succeeded in protecting carriers of lesser developed States by obtaining for them fair and equal opportunity to operate air services in routes that are shared by more established carriers of wealthier nations.[16]

Since GATS cannot sustain air transport services within a bilateral framework, it now remains to be seen whether the aviation community would move in the future toward placing air traffic rights in a multilateral or plurilateral system. In such an eventuality GATS would doubtless rejuvenate its efforts at seeking to include air transport services within its purview under liberalized market access and the MFN treatment clause. In this context, the role played by ICAO—the guardian and mentor of international civil aviation—becomes relevant.

ICAO has the mandate (under the Chicago Convention), experience and expertise in a wide range of air transport matters—technical, economic and legal. Issues of operating arrangements, market access, pricing and capacity for the designated airlines of each State are the subject of bilateral air transport agreements between States, except for arrangements within the European Economic Community for mutual relations between member States. International air transport is, in effect, conducted under an extensive network of some 3000 separate bilateral agreements or treaties. ICAO has taken the position that international air transport is an

[14] *Policy and Guidance Material on the Economic Regulation of International Air Transport*, Doc 9587, Second Edition, 1999, Appendix 2 at A2-2. See also *Manual on the Regulation of International Air Transport*, Doc 9626 First Edition 1996, Appendix 5, Guidance Material for Users of Air Transport, at A5-1.

[15] These conditions are the result of an agreement reached on 11 February 1946 by the United States and the United Kingdom in Bermuda. For a clear analysis of the Bermuda Agreement see de Murias (1989), pp. 52–72.

[16] See generally, Abeyratne (1993), p. 3.

economic activity in which there is a strong national interest and involvement as well as a long established, comprehensive and detailed structure of standards, principles and operating arrangements.

ICAO believes it important to draw to the attention of GATS and its member States certain critical features of international air transport which are relevant to any present or future consideration of how air transport should be treated in the context of the trade in services negotiations. The main consideration that impels ICAO to maintain steadfastly, its position as the guiding force behind air transport services is that it feels that bilateralism at the operating level has over the decades proved to be a flexible system which allows States to pursue their objectives, whether these be open and competitive or more protective or restrictive regimes for their airlines. ICAO strongly maintains that any external multilateral framework which sought general or limited application would need to recognize and be compatible with this existing structure of air transport.

Nevertheless, multilateralism in the form of a broad-based consensus on principles and guidance to States in the conduct of their air transport activities has enjoyed renewed interest in ICAO in recent years. While seeking to progressively develop positions and guidance to assist States in their regulatory/economic activities, ICAO recognizes the sovereignty of States in pursuing their own national air transport policies and objectives. ICAO's role in this sphere is therefore merely consultative and recommendatory without being incompatible with liberalization in this sector. ICAO has also expressed its resolve to continue to cooperate with GATT and the GNS in its trade in services discussions in order that ICAO's views and concerns and the particular features of the international air transport sector are properly taken into account by GATT and the GNS.

The Organization's position on the regulation of air transport services was formally adopted at its 7th Assembly held in June/July 1953 where Assembly Resolution A 7-15 resolved that there was no prospect at the time of achieving a universal multilateral agreement, although ICAO acknowledged that the achievement of multilateralism in commercial rights remained an objective of the Organization. This Resolution is still in force and reflects ICAO's commitment to achieving an acceptable multilateral basis for air transport services.

Later, at its 26th Session in September/October 1986, the ICAO Assembly adopted Resolution A 26-14, which reaffirmed that ICAO was the multilateral body in the United Nations system competent to deal with international air transport and urged contracting States which participated in any multilateral negotiations on trade in services where international air transport was included to ensure that their representatives were fully aware of potential conflicts with the existing legal system for the regulation of international air transport. The Resolution also requested the ICAO Council to actively promote a full understanding by international bodies involved with trade in services of the role of ICAO in international air transport and the existing structure of international agreements regarding air transport. This Resolution helped sensitize States and GATT regarding the air transport sector. Although this Resolution is no longer in force, it reflects adequately, ICAO's philosophy on the subject. In view of the significant recent and possible future developments in the trade in service negotiations, the question arises however, as to whether this policy is fully adequate to continue to serve the interests of ICAO and international air transport over the next few years or whether it requires reassessment and additional directives from the Assembly.

Assembly Resolution A26-14 gave guidance to States and the Council and expressed certain concerns but it did not set out an organizational view on the inclusion of international air transport in a multilateral agreement on trade in services. A future session of the Assembly may consider developing such a view for transmission to GATT and the GNS as well as to Contracting States.

One possible view the Assembly may consider might be that air transport should not be included in a services agreement. The adoption of such a position by ICAO could be grounded on two of the concerns found in Resolution A26-14. One is the Organization's concern about ICAO's role as the United Nations specialized agency responsible in air transport matters. The other is the Organization's concern for the integrity of the Chicago Convention principles and arrangements and the widespread system of bilateral air transport agreements that are a consequence of those principles.

Airlines are faced with the imminent prospect of the future realm of commercial aviation being controlled by a group of air carriers which may serve whole global regions and operated by a network of commercial and trade agreements. Regional carriers will be predominant, easing out niche carriers and small national carriers whose economics would be inadequate to compare their costs with the lower unit costs and joint ventures of a larger carrier. It is arguable that a perceived justification for open skies or unlimited liberalization exists even today in the bilateral air services agreement between two countries, were, *fair and equal opportunity to operate* air services is a *sine qua non* for both national carriers concerned. This has been re-interpreted to mean *fair and equal opportunity to compete* and later still, *fair and equal opportunity to effectively participate* in the international air transportation as agreed.[17] Of course, there has been no universal acceptance of this evolution in interpretation and carriers and States whose nationality such carriers bear, have maintained their own positions tendentiously.

ICAO has suggested the following preferential measures for the consideration and possible use of its member States who are at a competitive disadvantage when faced with the mega trends of commercial aviation and market access:

a) the asymmetric liberalization of market access in a bilateral air transport relationship to give an air carrier of a developing country: more cities to serve; fifth freedom traffic rights[18] on sectors which are otherwise not normally granted; flexibility to operate unilateral services on a given route for a certain period of time; and the right to serve greater capacity for an agreed period of time;

[17] Wassenbergh (1996), p. 80.

[18] The right to uplift or discharge passengers, mail and cargo in a country other than the grantor State.

b) more flexibility for air carriers of developing countries (than their counterparts in developed countries) in changing capacity between routes in a bilateral agreement situation; code-sharing to markets of interest to them; and changing gauge (aircraft types) without restrictions;
c) the allowance of trial periods for carriers of developing countries to operate on liberal air service arrangements for an agreed time;
d) gradual introduction by developing countries (in order to ensure participation by their carriers) to more liberal market access agreements for longer periods of time than developed countries air carriers;
e) use of liberalized arrangements at a quick pace by developing countries carriers;
f) waiver of nationality requirement for ownership of carriers of developing countries on a subjective basis;
g) allowance for carriers of developing countries to use more modern aircraft through the use of liberal leasing agreements;
h) preferential treatment in regard to slot allocations at airports; and
i) more liberal forms for carriers of developing countries in arrangements for ground handling at airports, conversion of currency at their foreign offices and employment of foreign personnel with specialized skills.[19]

These proposed preferential measures are calculated to give air carriers of developing countries a head start which would effectively ensure their continued participation in competition with other carriers for the operation of international air services. Furthermore, improved market access and operational flexibility are two benefits which are considered as direct corollaries to the measures proposed.

While the open skies policy sounds economically expedient, its implementation would undoubtedly phase out smaller carriers who are now offering competition in air transport and a larger spectrum of air transport to the consumer. Lower fares, different types of services and varied inflight service profiles are some of the features of the present system. It is desirable that a higher level of competitiveness prevails in the air transport industry, and to achieve this objective, preferential measures for carriers of developing countries would play a major role.

In addition, to addressing the preferential measures proposed by ICAO, which would be of immense assistance to carriers of developing countries if implemented, it would be prudent for the international aviation and trading community to consider the larger issue of funding, whereby long term low interest loans could be made available to carriers of developing countries through such institutions as the World Bank and the International Monetary Fund. Some consideration could also be given to a balanced distribution of aircraft throughout the world, whereby developing countries could have access to aircraft which have been discarded by

[19] See *Study on Preferential Measures for Developing Countries*, ICAO Doc AT-WP/1789, 22/8/96 at A-7–A-9. For a more recent revision of guidelines, see, *Policy and Guidance Material on the Economic Regulation of International Air Transport*, ICAO Doc, 9587, Second Edition, 1999, Appendix 3 at A3-1–A3-3.

their more affluent counterparts. An equitable system of leasing these aircraft is a possibility that could be considered in this regard.

The exemption of aircraft operated by carriers of developing countries from technological standards (to the extent possible) which may apply to modern aircraft is another useful tool which could be addressed under the umbrella of preferential measures. Aircraft engine emission standards and noise regulations are some examples which could be examined in this regard.[20]

Preferential measures may also be considered on a collective basis whereby air traffic rights could be used by a carrier of one country on behalf of another carrier representing another country. This would help, particularly in the event of a developing country not being able to launch its own airline or is unable to allocate its national carrier on a particular route due to economic reasons. This principle could also be extended to cover instances where airlines from developing countries could combine their operations by using their collective air traffic rights. For example, airlines of countries A and B who have been granted air traffic rights to operate air services from their countries to countries C and D respectively would be able to operate one joint service to countries C and D in one flight, using their collective traffic rights under this scheme.

It could be argued on behalf of the airlines that as far as possible, developing countries should be released from the obligation to own and control their air carriers or to have their carriers substantially owned and controlled by their nationals. It is only then that countries which cannot fully finance their carriers could maintain them and provide a well rounded competition in the air transport industry.

5.1.3 Legal Issues

The fundamental postulate of international trade lay in the General Agreement on Tariffs and Trade (GATT) system of 1947.[21] Under this agreement, in Article 1, a subsidy is defined, *inter alia*, as a benefit that is conferred by financial contributions by governments to a trading body which transfers funds such as grants, loans, equity infusion and potential transfer of loan guarantees. GATT 1947 has now given way to the WTO Agreement of 1995, Article 3 of which requires a member not to grant or maintain subsidies which are based on export performance. These subsidies, also called ‘export subsidies’, fall into the same category as subsidies that require as a condition of payment the use of domestic over imported goods for the purpose of the prohibition. Therefore, a member of the GATT system (now WTO) who

[20] For a detailed discussion of regulations on aircraft noise and engine emissions, see Abeyratne (1996) at chapter 3, pp. 271–313.

[21] BISD/IV, T.I.A.S No 1700, 55 U.N.T.S. 188. The GATT 1947 agreement is also contained in *Results of the Uruguay Round of Multilateral Trade Negotiations: The Legal Texts*, Geneva, GATT Secretariat, 1994.

considers that a subsidy granted by another member to a trading entity is unlawful, could engage in consultations with that member with a view to clarifying the latter's actions and facts of the case, and with a view to arriving at a mutually agreeable solution, or, as has happened in the culmination of the *Airbus/Boeing* dispute, refer the matter to the WTO dispute settlement mechanism.[22]

At first impression, principles on the law of subsidies as entrenched in international economic law are seemingly composed of a patchwork of provisions emerging from various agreements and amendments.[23] However, a deeper examination reveals that State subsidies are governed under a central theme which discourages unfair trade practices. Subsidies, which are government grants or bounties, are an integral part of international trade and entitle a government, by a selective process, to assist trading services and entities to the betterment of society. At the negotiating process when GATT was formed, subsidies were considered less of a trade obstacle than tariffs and quantitative restrictions. This was brought to bear in the 1948 Havana Charter of the International Trade Organization (ITO) which only contained a general ban on subsidies in its Article 26. The GATT of 1947, which replaced the ITO, did not prohibit subsidies.[24] It merely required that members reported to other members any subsidy which directly or indirectly affected exports by increasing exports or reduced imports into its territory.[25] This was specific to Article XVI:3 GATT 1947 which provided that if a contracting party (to the GATT) were to grant directly or indirectly any form of subsidy which operated to increase the export, the party concerned need only advise other parties of its action. However, in 1955, when GATT 1947 was reviewed, Article XVI was amended to prohibit export subsidies of non-primary products and to avoid the use of subsidies on the export of primary products.

Under the current WTO system, the Uruguay Round Subsidies Agreement, in Article 3 prohibits export subsidies to non-agricultural products.[26] With regard to internal subsidies, Article XVI:1 obligates members under the GATT 1947 system not to cause by means of any subsidy internally serious prejudice to competition.

It is worthy of note that in March 1982, a Panel was appointed by the WTO to address the validity of a claim by the United States that a German insurance programme for exchange rate risks incurred by Deutsche Airbus was unduly affecting competition between the USA and the European manufacturer, constituting an export subsidy contrary to the meaning and spirit of Article 9(f) of the Tokyo Round Subsidies Code.[27] There were allegedly two different texts applicable to the

[22] Article 4 of the WTO Agreement.

[23] Benitah (2001), p. 1.

[24] Rafiqul Islam (1999), p. 216.

[25] Article XVI of GATT 1947.

[26] In the Tokyo Round Subsidies Code, there was no formal definition of an export subsidy, although there were several examples cited.

[27] *German Exchange Rate Scheme for Deutsche Airbus*, Report of the Panel, March 4 1992. *See* 1992 WL 792947, SCM/142.

dispute (as alleged by Airbus) namely the Tokyo Round Subsidies Code and the Agreement on Trade in Civil Aircraft. However, the latter, in Article 6 provided that signatories shall take into account the special factors which apply in the aircraft sector, in particular the widespread governmental support in the area, together with the economic interests of State parties and the will of the world aircraft manufacturers to participate so that the world civil aircraft market is expanded. In the present dispute between the two manufacturers, it may well be, that if an arbitral panel were confronted with similar provisions in conflicting instruments, specific agreements entered into between the two States parties regarding assistance to their respective aircraft manufacturers will be taken into consideration over general subsidies provisions existing under the WTO umbrella.

5.1.3.1 Principles of Trade Practices and State Aid in Some Jurisdictions

The fundamental principle underlying State aid is that it disrupts normal competitive forces of the market if the State subsidizes certain firms and products to the detriment of others. Unsubsidized corporate entities could run out of business trying to compete with those receiving State aid, thus losing their right to compete fairly and equally. One of the corollaries to iniquitous State aid is the loss of employment for those in an unsubsidized environment. Article 87 of the EC Treaty allows State aid having a social character, granted to individual consumers as well as aid calculated to rectify damage caused by natural disasters and provide economic development of underdeveloped areas. State aid is also acceptable in instances where an important project of value to the European Union is assisted, including the promotion of culture and heritage conservation. The line is drawn when State aid tends to distort competition and certain firms are favoured. State aid in the above circumstances can comprise State grants; interest relief; tax relief; State guarantee or holding; and provision by the State of goods and services on preferential terms.

Article 81 of the EC Treaty prohibits such agreements as directly or indirectly fix purchase or selling prices or any other trading conditions; limit or control production, markets, technical development or investment; share markets or sources of supply; apply dissimilar conditions to equivalent transactions with other trading parties, thereby placing them at a competitive disadvantage; and make the conclusion of contracts subject to acceptance by other parties of supplementary obligations which, by their nature or according to commercial usage, have no connection with the subject of those contracts. These conditions are imposed on agreements between undertakings which are defined as independent entities performing some economic or commercial activity.

Article 82 provides that any abuse by one or more undertakings of a dominant position within the Common Market or in a substantial part of it shall be prohibited as incompatible with the Common Market insofar as it may affect trade between Member States. The Article prohibits: direct or indirect imposition of unfair purchase or selling prices or unfair trading conditions; limitation of production, markets or technical development to the prejudice of consumers; application of dissimilar conditions to equivalent transactions with other trading parties, thereby placing them at a competitive disadvantage; and conclusion of contracts subject to acceptance by the other parties of supplementary obligations which, by their nature or according to commercial usage, have no connection with the subject of such contracts.

Of some analogy, in principle, to the aircraft manufacturers 'subsidies issues' are the underlying principles of the *Ryanair* Case[28] which recognized that in Europe, there is entrenched legislation of the European Union concerning predatory pricing, competition and fair trade. As mentioned before, Article 82 of the *Treaty of Rome*, which prohibits abuse of dominant position by a carrier through a determination of the relevant market and market share enjoyed by the carrier under evaluation, was considered relevant to the *Ryanair* case before the European Court of Justice. The case concerned Article 85.1 (a) of the Treaty of Rome which governs Common Policies relating to Common Rules on Competition, Taxation and Approximation of Laws prohibits, as incompatible with the Common Market, *inter alia* all agreements between undertakings, which have as their objectives the preventive restriction or distortion of competition within the common market which directly or indirectly fix trading conditions. Within the broad rubric of this principle, the European Commission addressed information received that Belgium had granted a number of advantages to the airline Ryanair for the operation of air services at Charleroi Airport, on the basis that such advantages were inconsistent with Article 88(2) of the Treaty of Rome. During the proceedings, the Commission heard comments from interested parties.

The facts of the case were that on 6 November 2001, the owner of Charleroi Airport (Brussels South Charleroi Airport (BSCA) is a public sector company contracted by the Walloon Region, which has managed the airport under a 50-year concession agreement), the Walloon Region, signed an agreement with Ryanair giving the airline a reduction of approximately 50 % of the landing charge at the airport which had ordinarily been.

This information is matched with the pricing practices of the carrier which must not be lower than average variable costs. The *Competition Act* of 1998 of the United Kingdom links predatory pricing with dominant position and uses a process similar to that of the European Union[29] in assessing price–cost relationships. Germany has similar legislation in the *GesetzgegenWettbewerbsbeschränkungen* (GWB) which is the Act Against the Restraint of Competition, which identifies predatory practices as an abuse of dominant position if the predator is dominant in the market; the

[28] Abeyratne (2004), pp. 585–601.

[29] Both the EU and the United Kingdom uses the *AKSO NV* case as a benchmark where a Dutch chemical company, with a 65 % market share of its flour bleach product was found to be abusing its dominant position. The European Court of Justice found that price below average variable cost by means of which dominant competitor seeks to eliminate its competition is regarded as an abusive practice. *See AKZO Chemie BV* v. *EC* (1991) ECR 1-3359 at paras. 71–72.

conduct of predatory pricing is sustained and continuous and pricing is below average costs without objective justification.[30]

The United States competition law has as its genesis the *Sherman Act* of 1890 followed by the *Clayton Act* of 1914 (which was later amended in 1936). Such established legislation has been interpreted judicially to require two criteria: pricing must be below average variable costs and there has to be proof of recoupment of losses incurred during the alleged period of predatory pricing. In the 2001 case of *US* v. *AMR* Corp[31] the court held that an air carrier, which matches prices and increases output when faced with competition from low cost carriers, is not guilty of monopolization of the market even if the carrier fixed by the Walloon Government for all carriers serving the airport. The common rated charges base had been fixed by a decree issued in 1998 by the Government and was applicable to airport taxes that included landing, passenger and parking fees. *See* Decree of the Walloon Government of 16 July 1998 laying down the fees to be levied for the use of airports within the Walloon Region, *MoniteurBelge* of 15 September 1998. *See* 2004/393/EC, Commission Decision of 12 February 2004, reproduced in the *Official Journal of the European Union*, L 137/1 (30.4.2004) at p. 7. It was claimed that the reduction granted was discretionary on the part of the Walloon Minister of Transport, by means of a private contract by-passing the requisite statutory process. The deviation from established practice was, it was alleged, not in keeping with the required objectivity as the charges were calculated on the basis of each embarking passenger, instead of the usual tonnage weight of the aircraft. Furthermore, as an integral part of the agreement between the Walloon Region and Ryanair, the former had undertaken to compensate the latter for any loss of profit arising from a change in the level of airport taxes that might occur during the years 2001–2005 as a result of an internal decision taken by the Region. Under the agreement, the airport is authorized to keep 65 % of the fees collected by the Walloon Region from airlines serving the airport. The Belgian authorities had allowed this retention on the basis that the airport would undergo expenditures in welcoming, embarkation and disembarkation of passengers reverted to its original level of pricing after the low-cost carrier concerned had left the market. The court based its decision on the fact that the carrier had not priced its fare at a level below a appropriate level of cost. The carrier, in this instance, was found to be meeting the competition fairly rather than undercutting the other carrier, and there was no evidence that the carrier would recoup its losses through supra competitive pricing. Predation in Canada is brought within the purviews of both civil and criminal law where Section 50(1)(c) of the Canadian *Competition Act* recognizes selling at an unreasonably low price an act of predation when such practice is calculated to eliminate competition or lessen a competitor's ability to compete.

The Australian *Trade Practices Act* of 1974, which is administered through the Australian Competition and Consumer Commission, provides in Section 46 that,

[30] *GesetzgegenWettbewerbsbeschränkungen* (GWB) Section 20(4).

[31] U.S. District Court, District of Kansas, 27 April 2001, 28 Avi 15, 204.

when a firm takes control of dominant market power, particularly with intent to lessen or eliminate competition, the onus is on the person holding the position of dominance to prove his actions are not tantamount to predatory practices. The criterion used is that recoupment through pricing at supra competitive levels was a *sine qua non* to prove predatory pricing.

5.1.3.2 Issues at Stake

The issues before an arbitral panel of the WTO will be similar to earlier determinations, in particular the *Airbus* dispute discussed above. As in the earlier Airbus arbitration, the European Commission might not oppose the establishment of a Panel, but it might argue, as it did in that instance, that the case should be brought before the Committee on Trade in Civil Aircraft. The Committee was established under the aegis of the Agreement on Trade in Civil Aircraft which, like the Subsidies Code, was negotiated in the context of The Tokyo Round. The EC claimed that, the existence of the Agreement on Trade in Civil Aircraft was proof of the specificity of this kind of trade and thus of its legal regime. The EC also submitted that, should the Tokyo Round Subsidies Code be the only text referred to, this would prevent the examination of all aspects of the issue and would deprive the EC of its rights under the Agreement on Trade in Civil Aircraft.

The above notwithstanding, the President of the Committee on Subsidies and Countervailing Measures announced the failure of consultations concerning modified terms of reference, forcing the parties to a fall back on the 'standard' terms of reference envisaged in Article 18.1 of the Code, requiring the Panel to review the facts of the matter referred to the Committee by the United States in SCM/IO8 and, in light of such facts, shall present to the Committee its findings concerning the rights and obligations of the signatories party to the dispute under the relevant provisions of the General Agreement as interpreted and applied by the Tokyo Round Subsidies Code. The position taken by the EC was calculated to ensure that the Agreement on Trade in Civil Aircraft was included in the Panel's terms of reference, for the reason that although this Agreement recognized in Article 6.1 that the provisions of the Tokyo Round Subsidies Code applied to trade in civil aircraft, it also contained in the same Article the provision that signatories should also take into consideration the special factors which apply in the aircraft sector, in particular the widespread governmental support, their international economic interests, and the desire of producers of all Signatories to participate in the expansion of the world civil aircraft market.

In all probability, the above provision diffused the effect of the obligation arising from Article 9 of the Tokyo Round Subsidies Code when interpreted in the perspective context of trade in civil aircraft. From a legal context, the issue was thus a matter of conflict between the Agreement on Trade in Civil Aircraft and the Tokyo Round Subsidies Code in search of primacy. There are two different ways in which to view this dichotomy—a subjective approach and an objective approach. Whatever maybe the ideal choice of approach, the legal conclusion remains the same. Firstly, it is important to note that before discussing these approaches, one should note that the application of rules in the context of the GATT/1947 system was problematic. For one, the system was not endowed with a judicial body such as the International Court of Justice which has the authority to determine the relevant legal sources applicable to a given case and offer its opinion. However, in a Panel under the GATT/WTO system, in particular the fragmented GATT/1947 legal system, where, once the terms of reference of the Panel included exclusively, for example, the Tokyo Round Subsidies Code, the Panel could not refer to other sources, even if it wanted to do so. A reference to other sources would have been possible only in the context of explicit provisions linking the Tokyo Round Subsidies Code and the Agreement on Trade in Civil Aircraft. The absence of such explicit provisions was the precursor to the legal issue. It is also worthy of note that in the context of the GATT/1947 system, rules of international public law could, at best, play a role only in the preliminary phase where the Panel's terms of reference were negotiated. At this stage, the EC could well have invoked the applicability of rules of international public law, basing its action on the premise that the rules required special terms of reference in the *Airbus* case. However, it was incontrovertible that the Parties were constrained under a matter of a diplomatic nature which hinged on negotiation, comity and consequences arising out of possible non-conformity. Although one can assume that the case could have been better served under the judicial structure of the International Court of Justice, one should bear in mind that within the fragmented GATT/1947 legal system, arguments could only be made within a diplomatic umbrella within which the Panel's terms of reference were negotiated.

The subjective approach is based on the fact that in the light of two treaties seemingly applying to the same set of facts, the intention of the parties would be the relevant issue as a starting point. In this regard, what is important is that in both the Tokyo Round Subsidies Code and the Agreement on Trade in Civil Aircraft, there were no explicit provisions pointing to the intention of the signatories with regard to the primacy of one agreement over the other in a given case. It was thus prudent to look for reasons underlying an implied intention. In this respect, the EC's argument based on the specificity of trade in civil aircraft and its acknowledgment of a specifically applicable agreement was relevant. There is logic in the assumption that the existence of provisions relating to subsidies in the Agreement on Trade in Civil Aircraft brought to bear the existence of an implied intention of the signatories to establish a supplementary *Lex Specialis* to the more general provisions of the Tokyo Round Subsidies Code. The word 'supplementary', is used in the context of Article 6.1 of the Agreement on Trade in Civil Aircraft, which in fact recognized that the provisions of the Tokyo Round Subsidies Code applied to trade in civil aircraft.

The alternative approach—which was the objective approach—was irrevocably based on the fact that in any legal order, the international order takes prime place over domestic rules and is intended to resolve its own conflict of norms. With regard to the two counterpart agreements applicable to the *Airbus* case, they were not successive but contemporaneous and negotiated simultaneously in the Tokyo

Round. However, the Agreement on Trade in Civil Aircraft seemed to be a supplementary *lexspecialis* in the context of subsidies granted to civil aircraft. Ultimately, the EC prevented the adoption of the Panel's Report whose terms of reference only included the Tokyo Round Subsidies Code. This Report that was developed on the case came to the conclusion that the German programme contravened Article 9 of this Code and the *Airbus* case was resolved by means of a bilateral agreement between the EC and the United States.

The Airbus/Boeing subsidies issue came under the *Agreement on Subsidies and Countervailing Measures of the WTO* (SCM) which is enforced through the WTO dispute settlement system. The Agreement deals with two subjects closely related to each other. The first is concerned with issues of multilateral disciplines regulating the provision of subsidies and the second provides rules for the use of countervailing measures to offset injury caused by subsidized imports. Of these, the first will be applicable to the dispute between the two aircraft manufacturers in the determination of whether subsidies have in fact passed between the governmental authority and manufacturer concerned. The second would apply in determining what countervailing measures have to be taken by the WTO Dispute Settlement Body. Of primary concern would be Part 1 of the Agreement which specifies that the SCM would apply exclusively to subsidies that are specifically given to an enterprise or a group of enterprises or industries. The SCM goes on to define a subsidy as a provision which is composed of three fundamental elements: it has to be a financial contribution; it has to be made by a government or any public body within the territory of a WTO member; and it has to confer a benefit on the recipient. It must be noted that according to the SCM, subsidies, as defined, have to be specific. In other words, the subsidy has to be specifically provided to an enterprise or industry or group of enterprises or industries, resulting in the distortion of the allocation of resources within the economy.

The SCM recognizes four types of specificity: *enterprise specificity*, where a government identifies and provides a subsidy to a particular business entity; *industry specificity*, where a government targets a particular industrial sector; *regional specificity*, where a government targets business entities in a region or part of its territory; and *prohibited subsidies*, where a government subsidizes export goods or goods using domestic resources and inputs. Of these, clearly, the Airbus/Boeing dispute would be considered under the enterprise specificity. The next step would be to slot the dispute into one of two categories: prohibited category, and actionable category. The prohibited category is composed of two sub categories, the first being contingent upon export performance and the second being applicable to the use of domestic over imported goods. These two categories are prohibited on the ground that they are directly calculated to adversely affect trade, impacting the interests of other members. Actionable subsidies, on the other hand, are not prohibited, but nonetheless subject to challenge either through a multilateral dispute settlement system or through the imposition of countervailing actions. If adjudicated by the Dispute Settlement Body, the Airbus/Boeing dispute would clearly become an actionable issue under the latter category, on the basic assumption that the issue at stake is injury to a domestic industry.

The next issue would be to look at the dispute resolution process, which, in the WTO system essentially concerns inconsistencies in a commercial activity. Trade in civil aircraft and related services is a much evolved commercial process which stands today as one of the most prolific financial activities of the commercial world. It comes within the purview of the General Agreement for Tariffs and Trade (GATT) and is therefore now governed by the Marrakesh Declaration of 15 April 1994, which heralded the conclusion of the Uruguay Round of Multilateral Trade Negotiations and established the WTO.[32] The Marrakesh Agreement (hereafter sometimes referred to as the Agreement) ushered in a new era of global economic co-operation, reflecting the widespread desire to operate in a fairer and more open multilateral trading system for the benefit and welfare of its Member States.[33] A critical issue that emerged from this legal instrument was the function of the WTO as a forum for negotiations among its Members concerning their multilateral trade relations in matters which are dealt with under the agreements in the Annexes to the Agreement. The WTO may also provide a forum for further negotiations among its members concerning their multilateral trade relations, and a framework for the implementation of the results of such negotiations, as may be decided by the Ministerial Conference.[34] The WTO also administers the Understanding on Rules and Procedures Governing the Settlement of Disputes (hereafter referred to as the Dispute Settlement Understanding or DSU) in Annex 2 to the Agreement.[35]

Annex 2 to the Agreement, which addresses issues on dispute settlement and contains an Understanding on Rules and Procedures Governing the Settlement of Disputes, underscores the philosophy of dispute settlement reflected in the Agreement as being to strengthen the multilateral system of trade relations that the WTO represents. Article 23 of Annex 2 provides that when Members seek the redress of a violation of obligations or other nullification or impairment of benefits under the

[32] The WTO comprises a Ministerial Conference composed of representatives of all Member countries and meets at least once every 2 years. The Ministerial Conference carries out the functions of the WTO and takes necessary action to this effect. There is also within the WTO structure, a General Council composed of representatives of all Member countries and this Council meets as deemed appropriate. As necessary, the General Council conducts the functions of the Ministerial Conference when the latter does not meet. It also carries out its functions as assigned by the Agreement. Furthermore, the General Council also discharges the responsibilities of the Dispute Settlement Body.

[33] Marrakesh Declaration of 15 April 1994, para. 2. The Marrakesh Agreement came into being with the Marrakesh Declaration and is an integral part therefore of the Marrakesh Declaration. *See The Results of the Uruguay Round of Multilateral Trade Negotiations, The Legal Texts*. GATT Secretariat: Geneva 1994, for the text of the Declaration.

[34] Marrakesh Declaration, *id.* Article III.2. The Ministerial Conference is composed of representatives of all Members, which meets at least once every 2 years. The Conference carries out the functions of the WTO and takes action necessary to this effect. The Ministerial Conference also has the authority to take decisions on all matters under any multilateral trade agreements if so requested by a Member, in accordance with the specific requirements for decision-making in the Marrakesh Agreement and in the relevant multilateral trade agreement. *See* Article IV. 1. *Id.*

[35] *Id.* Article III.3.

covered agreements or an impediment to the attainment of any objective of the covered agreements, they shall have recourse to, and abide by, the rules and procedures of the Understanding. In such cases, members are required to abstain from making a determination to the effect that a violation has occurred, that benefits have been nullified or impaired or that the attainment of any objective of the covered agreements has been impeded, except through recourse to dispute settlement in accordance with the rules and procedures of the Understanding.[36]

Members are also required to follow the procedures set forth in Article 21—which call for prompt compliance with recommendations of the dispute settlement body, with particular attention being given to developing country Members with respect to measures which are subject to dispute settlement. This measure has been taken to determine the reasonable period of time for the Member concerned to implement the recommendations and rulings, and follow the procedures set forth in Article 22. Article 22 provides for international conditions for compensation and the suspension of concessions, in order to determine the level of suspension of concessions or other obligations and obtain authorization from the dispute resolution body before suspending concessions or other obligations in response to the failure of the Member concerned to implement the recommendations and rulings within that reasonable period of time.

The Dispute Settlement Body (DSB) which is established under the Council for Trade in Services, which are in turn established under the General Council of the WTO, was established to administer the rules and procedures in the Annex and, except as otherwise provided in a covered agreement, the consultation and dispute settlement provisions of the covered agreements. Accordingly, the DSB has the authority to establish panels, adopt panel and Appellate Body reports, maintain surveillance of implementation of rulings and recommendations, and authorise suspension of concessions and other obligations under the covered agreements.[37] The DSB is required to inform the relevant WTO Councils and Committees of any developments in disputes related to provisions of the respective covered agreements.[38] Under Article 3 of the Annex, Members affirm their adherence to the principles of the management of disputes heretofore applied under Articles XXII and XXIII of GATT 1947, and the rules and procedures as further elaborated and modified in the Annex.[39] This means that effectively, trade in civil aircraft and services related thereto would be governed by the initial structural format of the 1947 rules as amended. Emphasis is placed on the prompt settlement of situations in which a Member considers that any benefits accruing to it directly or indirectly under the covered agreements are being impaired by measures taken by another Member. The effective functioning of the WTO and the maintenance of a proper balance between the rights and obligations of Members has been identified as

---

[36] Annex 2, Article 23(a).

[37] Annex 2, Article 23(b).

[38] Annex 2, Article 2.1.

[39] Annex 2, Article 2.2.

critical in the dispute settlement process, which is required to be expedient.[40] Recommendations or rulings made by the DSB are calculated to aim at achieving a satisfactory settlement of the matter in accordance with the rights and obligations under the Understanding and under the covered agreements.[41]

One of the fundamental principles of the Understanding is that requests for conciliation and the use of the dispute resolution procedures should not be intended or considered as contentious acts. If a dispute arises, all Members are required to be involved in the procedures in good faith in an effort to resolve the dispute. Furthermore, complaints and counter-complaints in regard to distinct matters should not be linked.[42]

In the Airbus/Boeing dispute, as in any other, the WTO dispute settlement structure encouraged conciliation and mediation under procedures that are undertaken voluntarily if the parties to the dispute so agree.[43] Any party to a dispute may, at any time, request good offices, conciliation and mediation. They may begin at any time and be terminated at any time. Once procedures for good offices, conciliation or mediation are terminated, a complaining party may then proceed toward the establishment of a panel. At the request of the complaining party, a panel is established under the Agreement latest at the DSB meeting following that at which the request first appears as an item on the agenda of the DSB, unless of course, the DSB decides not to establish a panel at its meeting.[44] A Panel is composed of well-qualified governmental and/or non-governmental individuals, including persons who have served on or presented a case to a panel; served as a representative of a Member or of a contracting party to GATT 1947; or, as a representative to the Council or committee of any covered agreement or its predecessor agreement, or in the Secretariat; has taught or published on international trade law or policy; or, served as a senior trade policy official or member[45] and they are selected in such a way as to ensure the independence of the members and to give the panel a diverse selection of individuals with a wide spectrum of experience and exposure.[46] To this end, the WTO Secretariat maintains an indicative list of governmental and non-governmental individuals possessing the qualifications to assist in the selection of panellists as required.[47]

---

[40] Annex 2, Article 3.2.

[41] Annex 2, Article 3.3.

[42] Annex 2, Article 3.4.

[43] Annex 2, Article 3.10.

[44] Annex 2, Article 5.1.

[45] Annex 2, Article 6.1.

[46] Annex 2, Article 8.1.

[47] Annex 2, Article 8.2.

5.1.4 Past, Present and the Future

The future of air transport, in terms of demand forecasts, has already been predicted and it is an interesting mosaic of development and economic growth. The Report of an OECD[48] study[49] conducted between 1980 and 2013, notes that the airline sector has grown substantially and quotes ICAO's figures—that passenger trips increased from 4.028 billion in 1980 to 19.125 billion in 2012. International scheduled passenger traffic grew by 5.2 % in 2013 in comparison to 2012 and is expected to reach over 6.4 billion passenger by 2030.[50] ICAO forecasts that by 2030, an average annual growth rate of 4.5 % is forecast for passenger traffic (of both scheduled and unscheduled services) but growth during 2020–2030 will slightly decrease as markets mature.[51]

The OECD Report states that by 2026, it is expected that air transport will contribute $1 trillion to world's GDP.[52] *Airbus Industrie*, in its global market forecast (2014–2030)[53] identifies the key drivers of air transport as economic growth; increasing urbanisation; expanding middle class; and rise in migration, tourism and international students. The Airbus forecast predicts that emerging countries-regions (Asia and the Pacific, Africa, Middle East and South America) will overtake the developed countries-regions in terms of economic growth with a 10 % increase in growth in passenger travel. One of the notable drivers will be that the middle class in Asia Pacific countries will grow four times as that of the developed countries and international tourist arrivals will reach 1.6 billion in 2020, well above the 1.1 billion as of 2013.[54] *Airbus Industrie* has also forecast that between 2009 and 2028 there will be a demand for 24,951 passenger and freighter aircraft worth US $3.1 trillion, and that, by 2028 there will be 32,000 aircraft in service compared with 15,750 in 2009.[55]

---

[48] Organisation for Economic Co-operation and Development (OECD), established in 1961, promotes policies that are calculated to improve the economic and social well-being of people around the world. The OECD provides a forum in which governments can work together to share experiences and seek solutions to common problems. The Organization works with governments to understand what drives economic, social and environmental change.

[49] AIRLINE COMPETITION—Background Paper by the Secretariat, Directorate for Financial and Enterprise Affairs Competition Committee, 18–19 June 2014, DAF/COMP(2014)14.

[50] ICAO Press Release, 16 December 2013. The OECD Report also notes that the number of travellers has increased because, among many other things, prices have decreased significantly in response to increasing competition in the air transport market. For example, in 1974 the cheapest round-trip New York-Los Angeles flight (in inflation-adjusted dollars) that regulators would allow: $1442. Today one can fly that same route for $268.

[51] Global Air Transport Outlook to 2030, *Circ.333, AT/190*:2012 at 59.

[52] AIRLINE COMPETITION, *Supra* note 49 (this chapter), *Background Note* at 3. The Report goes on to say that worldwide, aviation and related tourism generate over 56 million jobs, of which 8.36 million are directly linked to the aviation sector. Around 35 % of international tourists travel by air.

[53] Global Market Forecast: Flying on Demand 2014–2033 at 16.

[54] *Id.* 17.

[55] Flying Smart, Thinking Big, *Airbus Industrie's Global Market Forecast, 2009–2028* at 16.

Boeing—the other large aircraft manufacturer, forecasts that the demand for new aircraft in the long term calls for 36,770 new airplanes, valued at $5.2 trillion. According to *Boeing*'s assessment, 15,500 of these airplanes (42 % of all new deliveries) will replace older, less efficient airplanes. The remaining 21,270 airplanes will essentially be for the aircraft capacity needed for be for fleet growth, particularly to stimulate the market in emerging economies and the development of innovative airline business models. Single-aisle airplanes continue to command the largest share of the market. For shorter-haul, point to point air transport, particularly operated by low cost carriers, *Boeing* forecasts that approximately 25,680 new single-aisle airplanes will be needed over the next 20 years. Another feature for the future is the extra large wide body aircraft which will further exploit current long routes as well as open with 8600 new airplanes.

However encouraging these figures might be, it is incontrovertible that, as an industry, air transport has wobbled its way through over the years with periods of unprofitability followed cyclically by bare profitability, largely due to the inability of airlines to adjust to vacillating traffic demands and peaks and troughs of its perishable product. There are other contributory factors to this phenomenon, such as the in ability of airlines to cut capacity when traffic demands are low; rigid contractual obligations to their employees and union action; discounted pricing by competitors; increasing competition exacerbated by disadvantages in reservations caused by websites which make pricing transparent; and the inevitability of operating from the same airports over sustained periods of time. Arguably the most ominous costs are rising labour costs and escalating fuel costs.[56]

5.1.5 The UAE Carriers

There are some airlines that have weathered the storm and become profitable and forward thinking. Although their business model may not be for all airlines, it is interesting to look into one particular airline which has forged ahead ever since its inception in 1985, where it was launched with a starting capital of US $10 million, 10 employees and 2 leased aircraft—a leased Boeing 737 and an Airbus 300-B4. Emirates Airline has now a fleet of more than 230 aircraft, and currently flies to over 140 destinations in more than 80 countries around the world, and the airline's network is expanding constantly. Over 1500 Emirates flights depart Dubai each week on their way to destinations on six continents. In recent years, Emirates has made numerous significant announcements regarding the future of its already state-of-the-art fleet. Emirates has on order more than 280 aircraft, with a total value of approximately US $138 billion as of November 2014. In the financial year 2013/2014, Emirates carried 44.5 million passengers and 2.25 million tonnes of cargo.[57]

[56] See ICAO Circular 333, *Global air Transport outlook to 2030 and Trends to 2040* at 44.

[57] See https://mobile.emirates.com/MobileAboutEmirates/global/english/emirates_story.xhtml.

Although United States carriers have accused Emirates (and some other carriers) of unfair competitive practices, alleging that, buttressed by subsidies given to the airline by the government of the United Arab Emirates,[58] Emirates is displacing the US carriers in their own market. This claim has been vehemently refuted with a counter argument that Emirates in fact gives money to the UAE government from profits made.[59] *Oxford Economics* stated in a 2011 report[60] that

> The success of Dubai's aviation sector derives from a number of strengths, which are the product of strategic decisions that the government of Dubai and the aviation sector have taken in the past. These strengths include: an awareness of aviation's economic importance on the part of the government of Dubai; openness; a consensus based approach to investment; a focus on growth and linking underserved markets; and efficient operations. Added to these is Dubai's favourable location at the intersection of Europe, Asia and Africa.[61]

Etihad, another UAE carrier which operates from Abu Dhabi, has also been confronted with the same accusation as has Emirates. In a study commissioned by *Etihad* it has been revealed that the accusation that these carriers "displaced" US carriers is tenuous as the real criterion should be end to end traffic and not carry over 5th freedom traffic through the UAE hubs. In that context it has been concluded that with regard to a passenger who flies from New York to Mumbai, there are various carriers offering services through hubs, making the competition open and fair.[62]

Despite glowing figures of future growth and aircraft purchases, we have come full circle in air transport where the phenomenon of protectionism of the early eighties in the last century has come back to haunt us. It is submitted that this problem started with the carriers and it should end with the carriers themselves. On a basic analysis, the American and European carriers who lag behind the emerging economies' carriers have to realise that the issue hinges on competition and not protectionism. The air transport industry stands at the crossroads of two major

---

[58] See http://pallone.house.gov/press-release/pallone-and-colleagues-lead-letter-opposition-abuse-open-skies-agreement.

[59] See ***No evidence Gulf carriers get 'artificial support', says Willie Walsh*** http://www.arabianbusiness.com/no-evidence-gulf-carriers-get-artificial-support-says-willie-walsh-564516.html#.VBUzWsWSxMg Also, Interview: Maurice Flanagan, Emirates Airline, by Shane McGinley, July 27 2010. See http://www.arabiansupplychain.com/article-4587-interview-maurice-flanagan-emirates-airline/2/.

[60] *Explaining Dubai's Aviation Model*, Oxford Economics, June 2011. See http://www.oxfordeconomics.com/my-oxford/projects/128910.

[61] *Id.* 5.

[62] The following choices are available to a passenger travelling from New York to Mumbai. American Airlines/British Airways with a connection in *London*; Emirates with a connection in *Dubai*; United/Lufthansa/Swiss Air with a connection in *Frankfurt, Munich, or Zurich*; Air India with a connection in *Delhi*; Turkish Airlines with a connection in *Istanbul*; Qatar Airways with a connection in *Doha*; or Delta/Air France/KLM with a connection in *Paris or Amsterdam.* See Empirical Investigation and Analysis of Economic Issues Raised in Restoring Open Skies: The Need to Address Subsidized Competition from State Owned Airlines in Qatar and the U.A.E., Edgeworth Economics, May 21, 2015 at 5.

influences—globalization and the information revolution—which have revolutionized the trading world by driving competition. The fact that the UAE carriers (as well as Qatar Airways and Turkish Airlines) have the geographic advantage of being in the centre of the long air routes of the world should not be taken as an argument supporting competitive distortion or disadvantage to other carriers which would in turn unduly curb the operations of the four carriers. The Air Transport Association (IATA)[63] has recognized that US $4.2 trillion is needed over the next two decades to properly offer connectivity to a travelling public who are creating an exponential demand for air transport.[64]

IATA's Director General Tony Tyler has said that what is needed is smart regulation from governments around the world in order to maximize the economic benefits of connectivity—jobs and growth. Tyler says that unfortunately, high taxation and poorly designed regulation in many jurisdictions make it difficult for airlines to develop connectivity. One of the problems identified by Tyler is that in addition to cost issues, airlines also face a hyper-fragmented industry structure owing to government policies that discourage cross-border consolidation. He suggests fresh thinking on all accounts.[65]

The American carriers have to face facts. In its latest survey, *Skytrax* ranks Qatar Airways as a 5 star airline[66] and Emirates as a 4 star airline. No US carrier is placed under these categories. Obviously, the fragmentation and divide and rule practices must give way to a more forward looking approach to air transport. As one commentator said:

> The liberalization of markets, the construction of a globalized economy and the spread of prosperity are defining legacies of the era of Western primacy. The fundamentals of this order are firmly in place, anchored by institutions like the World Bank and the World Trade Organization. But the maintenance of this order faces significant challenges. Due to the West's political and economic troubles, the Atlantic democracies may no longer be up to minding the store. The United States already seems to have lost its traditional enthusiasm for being the engine behind the global liberalization of trade.[67]

---

[63] The International Air Transport Association is a trade association of the world's airlines. These 250 airlines, primarily major carriers, carry approximately 84 % of total available air traffic.

[64] Schaal (2015). See http://skift.com/2013/07/01/the-airline-business-is-a-terrible-one-says-leading-airline-industry-group/.

[65] *Ibid.*

[66] A key factor behind 5-Star Airline Rating is the airline's ability to deliver a truly consistent and high quality of product and service. A great product is not by itself the key to a 5-Star Airline rating, and *Skytrax* Star Rating also places great emphasis on the quality of front-line service an airline provides. A 5-Star Airline rating recognises airlines whose front-line staff (across airport and onboard experience) deliver a true and consistent 5-Star standard of service delivery. See Emirates http://www.airlinequality.com/StarRanking/4star.htm.

[67] Kupchan (2012), p. 198.

5.1.6 The Mathematical Theories

There are three theories that are applicable to competition in air transport that would go towards helping carriers compete with each other. Jordan Ellenberg, a professor of mathematics at the University of Wisconsin-Madison, in his book *How Not to Be Wrong*[68] explains how one can go wrong if one does not follow mathematical logic in the reasoning and decision making process. The book is about the proper use of probability and statistics and how to reject counterintuitive precincts of mathematical thinking. This approach would apply almost to any discipline or practice, from running a business to politics.

There are seemingly three theories that lend themselves to the logic behind success at a business. One is the *Probability Theory*. *Encyclopaedia Britannica* identifies the Probability Theory as: "a branch of mathematics concerned with the analysis of random phenomena. The outcome of a random event cannot be determined before it occurs, but it may be any one of several possible outcomes. The actual outcome is considered to be determined by chance".[69]

The second theory is the *Game Theory*, which is a philosophy drawn on the discipline of applied mathematics that could be applied to politics and economics. *Investopedia* defines the Game Theory as: "the process of modeling the strategic interaction between two or more players in a situation containing set rules and outcomes".[70] The Game Theory—a quantum theory on anticipatory intelligence—is about maximising returns based on the strategic decisions to be made by contestants at economics, trade or politics. In air transport, the theory would help analyze interactions of carriers and strategies between them, thus enabling the airlines competing with each other to study strategic interactions between them. The outcome is a formal modelling approach to economic situations in which decision makers interact with other decision makers.

The third theory is called *Disruptive Innovation*, a business concept which is an innovation that helps create a new market and value network that disrupts the existing market. The theory of disruptive innovation was first coined by Harvard professor Clayton M. Christensen in his research on the disk-drive industry and later popularized by his book *The Innovator's Dilemma*, published in 1997. Examples of disruptive innovation abound in the commercial world. For instance, Wikipedia disrupted the market established for more than 200 years by *Encyclopedia Britannica*. The *IPhone* disrupted the market of the desktop computer and the laptop computer. A good example in the air transport industry which replaced legacy carriers in certain segments and routes, purely by the use of a new product that was more cost effective and efficient, is the low cost carrier.

As for the *Game Theory*, the carriers which consider themselves displaced could well apply their anticipatory intelligence to their opponents. The opposition could

[68] Ellenberg (2014).

[69] http://www.britannica.com/EBchecked/topic/477530/probability-theory.

[70] http://www.investopedia.com/terms/g/gametheory.asp.

also do likewise in anticipating and countering the strategies of its opponent. However, most importantly, disruptive innovation could play its part with a new tool that introduces an "economic market" with a new value network and cluster that could disrupt a repetitive economic environment that deprives the public of its needs for connectivity.

5.2 Market Economics and Air Transport[71]

The need for defragmenting air transport with a systemic approach should be considered in the context of conflicting commentaries. The 2011 Report of the World Bank stated:

> The world economy has entered a very difficult phase characterized by significant downside risks and fragility. The financial turmoil generated by the intensification of the fiscal crisis in Europe has spread to both developing and high-income countries, and is generating significant headwinds. Capital flows to developing countries have declined by almost half as compared with last year, Europe appears to have entered recession, and growth in several major developing countries (Brazil, India, and to a lesser extent Russia, South Africa and Turkey) has slowed partly in reaction to domestic policy tightening. As a result, and despite relatively strong activity in the United States and Japan, global growth and world trade have slowed sharply.[72]

The International Monetary Fund in December 2014 confirmed that this trend of slow growth in global trade has continued, stating that trade has been unexpectedly slow moving even after recovering in 2010 from the historic low recession of 2008–2009.[73] The IMF draws a distinction between the trade slump and income

[71] A few paragraphs at the commencement of this section has been taken from the author's book *Administering the Skies: Facing The Challenges of Market Economics*, Aracne Editrice: Rome 2012.

[72] http://siteresources.worldbank.org/INTPROSPECTS/Resources/334934-1322593305595/8287139-1326374900917/GEP_January_2012a_FullReport_FINAL.pdf. The Report also stated that the global economy is now expected to expand 2.5 and 3.1 % in 2012 and 2013 (3.4 and 4.0 % when calculated using purchasing power parity weights), versus the 3.6 % projected in June for both years. High-income country growth is now expected to come in at 1.4 % in 2012 (−0.3 % for Euro Area countries, and 2.1 % for the remainder) and 2.0 % in 2013, versus June forecasts of 2.7 and 2.6 % for 2012 and 2013 respectively. Developing country growth has been revised down to 5.4 and 6.0 % versus 6.2 and 6.3 % in the June projections. Reflecting the growth slowdown, world trade, which expanded by an estimated 6.6 % in 2011, will grow only 4.7 % in 2012, before strengthening to 6.8 % in 2013.

[73] Constantinescu et al. (2014). See http://www.imf.org/external/pubs/ft/fandd/2014/12/constant.htm. The IMF states that Trade grew by no more than 3 % in 2012 and 2013, compared with the pre-crisis average of 7.1 % (1987–2007). For the first time in over four decades, trade has grown more slower than the global economy. One commentator says: "In part, the weakness in world trade has clearly been due to the cyclical slowdowns in the Eurozone in 2012, and in the BRICs in 2013. Because trade between the member states inside the Eurozone is counted as part of world trade, the euro crisis has had a particularly large effect. But this effect should have been reversing

growth, where the latter has shown a distinct increase while trade has slowed down. This phenomenon is explained by the fact that although global income grew, trade has not been responsive to the growth. Some of the probable factors attributed to this slump are: changes in the structure of trade associated with the expansion or contraction of global supply chains (particularly between the United States and China); changes in the composition of world trade, such as the relative importance of goods versus services; changes in the composition of world income, such as the relative importance of investment and consumption; and changes in the trade regime, including the rise of protectionism leading to the fragmentation of the global marketplace.

Encouragingly, this gloomy picture will not stay for long. Recent forecasts of the IMF reflect 5.6 % growth in world trade and 4.1 % real GDP growth for 2019. It is reported that trade grew by 3.3 % in 2014, which is an increase from 2.7 % in 2013 and 2.1 % in 2012. However, it is still well below the long term average of growth of 5 %.[74] Such figures portend a recovery of global trade where it would exceed GDP growth. One commentator is of the view that global trade should benefit in particular from the projected pick-up in investment in advanced economies.[75]

This upward trend is consistent with the considered opinion of Fareed Zakaria who says:

> Over the last two decades, about two billion people have entered the world of markets and trade – a world that was, until recently, the province of a small club of western countries. The expansion was spurred by the movement of western capital to Asia and across the globe. As a result, between 1990 and 2010, the global economy grew from $22.1 trillion to $62 trillion, and global trade increased by 270 %.[76]

Against this scenario, Ruchir Sharma says of the future:

> Over the next decade, growth in the United States, Europe and Japan is likely to slow by a full %age point compared to the post World war II average, owing to the large debt overhang, but that slowdown will pale in comparison to a 3 to 4 %age average slowdown in China.
>
> The big story will be that China is too big and too middle-aged to grow so fast. And as it starts buying less from other emerging nations, the average pace of growth in emerging markets is likely to slow from nearly 7 % over the past decade to the 1950s and 1960s pace of around 5 %.[77]

Added to this factor is the contagion effect the Euro debt crisis was having on China and India where the decrease in demand for their goods was being felt

---

by the middle of 2013, so it is disturbing to see that the decelerating trend in trade flows has continued". See Davies (2013).

[74] Weldon (2015). See http://www.bbc.com/news/explainers-31661078.

[75] Boz (2014).

[76] Zakaria (2011), p. 21. The author goes on to say: "The so called emerging markets have accounted for over half of this global growth, and they now account for over 47 % of the world economy measured at purchasing power parity (of over 33 % at market exchange rates)." *Ibid.*

[77] Sharma (2012), p. 242.

worldwide and the effects were being felt as far as Australia which is the only G-20[78] economy that was not affected by the recession of 2008.

The woes of the air transport industry have always centered on the claim that aviation is effectively precluded from entering the free market due to government meddling. This seems to go contrary to the Keynesian view of the advantages of some degree of government control of the free market economy. Harsh restrictions on ownership and control of airlines and market access are stringent tools of government control. The argument for deregulation, as against government control, pervaded the economic crisis of 2008 where some argued that the bubble burst because the markets had not been free enough from government meddling and that the US Congress pushed lenders to have unfettered ability to lend to those who could not pay back their loans, leading to a housing bubble. Pankaj Ghemavat, referring to the 2008 bubble says:

> Media magnate Rupert Murdoch blamed the Government for the debacle, stating: "It's very easy to blame the free market but how did we get to the housing bubble? We got it because of Congress pushing Fannie Mae and Freddie Mac into lending money to people who couldn't afford it and blowing up the price of housing. . . .[79]

In the air transport sector this principle is reversed in that governments meddle in fettering the freedom of air carriers by imposing political and economic restrictions. This trend seems to continue, much to the disappointment of the air transport industry. Anne-Marie Slaughter of Princeton University speaks of a new world order where the State will not be the only actor in the international system but will still be the most important actor; the State will not disappear but will disaggregate into its component institutions, which will increasingly interact principally with their foreign counterparts across borders; these institutions will represent distinct national or State interests, even as they also recognize common professional identities; and government networks will exist alongside and sometimes within more traditional international Organizations.[80] For air transport this is good news in that the State will not be the only actor in the international system, which hopefully mean that International Organizations may be more empowered and that government networks may exist within international Organizations, promoting interest groups within such Organizations to push liberalization of air transport, the primary such Organization being ICAO.

[78] The Group of Twenty Finance Ministers and Central Bank Governors (also known as the G-20, G20, and Group of Twenty) is a group of finance ministers and central bank governors from 20 major economies: 19 countries plus the European Union, which is represented by the President of the European Council and by the European Central Bank Their heads of government or heads of state have also periodically conferred at summits since their initial meeting in 2008. Collectively, the G-20 economies account for more than 80 % of the gross world product (GWP), 80 % of world trade (including EU intra-trade), and two-thirds of the world population. They furthermore account for 84.1 % and 82.2 % of the world's economic growth by nominal GDP and GDP (PPP) respectively from the years 2010 to 2016, according to the International Monetary Fund.

[79] Ghemawat (2011), p. 12.

[80] Slaughter (2004), p. 18.

The answer to the economic slump lies in global growth, and growth comes from innovation. One thing we should be very clear of is that innovation and the "predict and provide" philosophy are antithetical and the former is the only way air transport can meet growing demands of the next two decades. On this, Nobel Laureate Michael Spence has said:

> Innovation, which is sometimes called technological progress, increases the production potential of an economy over time. That means that with the same amounts of capital, labor, raw material and energy, you can produce more – or more valuable – output. You can also think of it as reducing the cost of producing a given amount of output.[81]

Spence's argument is where the rub is for air transport. The answer that would ensure a correct approach to air transport lies basically in competition, but it does not stop there. As the UAE carriers have shown, and as reflected in the *OXFAM* report already cited, many factors come into play in winning at competition. The foremost is talent and creativity in applying disruptive innovation. As mentioned earlier, the *Edgeworth Report* shows that where it comes to fifth freedom traffic on a long thin route, the consumer has several choices. Emirates offers the most attractive connection. Emirates' success has come about through a holistic approach involving a flavorful combination of the factors influencing the entire airline product—from airport comfort to in-flight service superiority. This, as Spence says is innovation.

Klaus Schwab, the founder of the World Aviation Forum says:

> When financial policymakers attempt to promote economic growth, they almost invariably focus on looking for new ways to unleash capital. But, although this approach may have worked in the past, it risks giving short shrift to the role that talent plays in generating and realizing the ideas that make growth possible. Indeed, in a future of rapid technological change and widespread automation, the determining factor – or crippling limit – to innovation, competiveness, and growth is less likely to be the availability of capital than the existence of a skilled workforce.[82]

Schwab believes that instead of curbing disruptive innovation, governments should legislate towards encouraging their sustained growth, thus promoting competition and generating employment growth. This approach is incontrovertibly consistent with the inherent purpose of air transport which is to connect the world. Air transport is not primarily for the airlines or the States. It is first and foremost for the consumer. It unites the people of the world (as the Chicago Convention says) and prescribes ICAO's aim as per the Convention as "meeting the needs of the people of the world for safe, regular, efficient and economical air transport".[83]

---

[81] Spence (2011), p. 36.

[82] Schwab (2015).

[83] Chicago Convention, *supra*, Chap. 1, note 5, Article 44 d).

5.3 Competition and Innovation

A common misconception is that innovation should decrease with increased competition. In the discussion above, Klaus Schwab put to rest this erroneous assumption. Competition and innovation are mutually endogenous. The link between the two has been summed up in a paper published by Harvard University in 2014:

> First, an increase in competition leads to a significant increase in R&D investments by neck-and-neck firms. Second, an increase in competition decreases R&D investments by laggard firms. Moreover, this Schumpeterian effect[84] is significantly stronger the shorter the time horizon. Third, increased competition affects industry composition by reducing the fraction of neck-and-neck sectors, and overall, competition increases aggregate innovation.[85]

Thus in highly competitive markets such as in any high yield air transport market, laggard firms would tend not to emphasize on innovation. Economic theory suggests that the higher the level of competition the more beneficial it is to the economy and society. This is because competition inevitably leads to more efficient allocation of resources based on economies of scale. Higher efficiency leads to the creative destruction of the less efficient competitor. At the same time, government regulation is necessary to promote a high level of competition and establish policy to that effect. Improved technology, innovation and creativity promote competition.

The trouble with air transport has always been that, while on the one hand, the preamble to the Chicago Convention speaks of international air transport services being established on the basis of equality of opportunity, yet Article 6 of the Convention provides that a scheduled air transport service may be operated only under permission of the State being flown into. Curiously, it is this very dichotomy that lends itself to airlines being attracted to disruptive innovation, when States can be overtly protective of their national carriers. An example is *Ryan Air* which broke into the nascent low cost carrier model and disrupted the existing market developed by the legacy carriers of Europe.[86] *Southwest Airlines* went further, by disrupting not only the air transport market but also the road transport market. From its inception, Southwest Airlines aggressively competed against the bus transport in Texas. The increased customer-base was attracted to the low fares of the airline, as well as the connectivity and the increased number of flights that offered the speed inherent in air transport over surface transport.[87]

---

[84] The Schumpeter effect is associated with the notion of creative destruction put forward in the 1930s by the celebrated economist Joseph Schumpeter, which introduced the process by which new innovations replace older technologies. Start-up airlines have to be mindful of being overrun by more established airlines, making creative destruction a common phenomenon in air transport.

[85] Aghion et al. (2014). See scholar.harvard.edu/files/aghion/files/causal_effects_of_competition.pdf.

[86] Low cost carriers emerged primarily because of the globalization of the air transport industry and market deregulation in most parts of Asia and Europe, which encouraged new enterprises to approach the air transport market with vigour and energy. See Lawton (2002), p. 1.

[87] Defining Performance In Disruptive Innovation, 15 June 2009. See *Forbes Magazine* at http://www.forbes.com/2009/06/15/southwest-apex-learning-personal-finance-clayton-christensen-innovation-airlines.html.

Disruptive innovation in the air transport industry is based on two strategies: service strategy and pricing strategy. These two combined justify the three basic features of a successful business strategy which displaces an existing market: availability of goods and services; good price and quality; and value for money. When these practices are applied to the airline industry, one finds that an established carrier is much more vulnerable to disruptive innovation than most other industries. Christensen—the Harvard academic who coined the term said:

> Companies in other industries have more options when attacked. Computer companies, for instance, can migrate to high-end products with lower volumes…the high fixed-cost structure of hub-and-spoke airlines means they can't run away from the volume in the lower tiers of the market.[88]

Whilst one cannot question the irrefutable logic of this statement, disruptive innovation does not always carry a guarantee of business success. The instance of *Skybus* is a case in point. *Skybus*—a start up US low cost carrier attempted to enter an existing market but folded up after a short while because it did not have a strategic business model with which it could compete with the existing market. *Skybus* did not have the approach of the more successful carriers such as *Southwest Airlines* and *RyanAir*:

> *Southwest* succeeded because it coupled the most attractive secondary routes with a very low-cost business model. *RyanAir* has been able to succeed because its cost structure is incredibly low, its operations are incredibly efficient, and it has high capacity utilization. In both cases disruptive success traces back to business models that allow prosperity at low price points.[89]

Skybus, in trying to create its new business, could have gone in one of two directions—by building sustaining innovation i.e. taking the existing market and improving on it or take on a competitor with disruptive innovations that either create new markets or take root among an incumbent's worst customers. Another approach could have been to exploit the low end of the market by taking customers away from the established market by offering more attractive terms. In both these approaches the business planning process would be crucial. There should be no haste in looking for growth but focus should rather be on profitability. A disruptive innovation in air transport should take into account the key drivers that would determine success in the air transport business. These are: improved passenger experience where the product displaced does not have as good a service than the replacing airline; more efficient revenue generation where costs are minimized and value of the product is enhanced, making the customer the focus; more advanced technology than the competitor who is being displaced; loyalty and benefits to the consumer; and efficiency in financial transactions which make it easier for the consumer to purchase the airline product; and the ability to adapt to changing circumstances.

---

[88] Disruption: Flying the Not-So-Friendly Skies, *Working Knowledge for Business Leaders*, Harvard Business School. See http://hbswk.hbs.edu/archive/3736.html.

[89] Anthony (2008), see https://hbr.org/2008/04/why-its-so-hard-to-disrupt-the/.

It is our intelligence and pioneering spirit that have brought us to where we are. At the same time, it is our innate reluctance to change that has kept us back from more progress. In between these two are our indomitable spirit and unfailing belief in ourselves, that we can progress from baby steps to full grown adulthood in any field we choose, which keep us going.

When the Wright brothers invented and built the world's first successful airplane and made the first controlled, powered and sustained heavier-than-air human flight on December 17, 1903 in Kitty Hawk, North Carolina, none would have thought that a little more than a hundred years later, that initial flight would have evolved into a super jumbo jetliner that could carry at least 550 people across the world. The Wright brothers went step by step to achieve their feat. In July/August 1899, they built, and one of the brothers—Wilbur—flew a biplane kite in order to test the "wing-warping" method of controlling a flying machine. This experiment encouraged the Wrights to proceed with constructing a flying machine with a pilot.

This story of aviation stands as a monument to air transport in today's context. Innovation and creativity stand as the cornerstone of disruptive business practices that create new markets and build successful air transport enterprises. On the one hand is Emirates Airline as a mega carrier and on the other hand is Air Asia—a low cost carrier which was named the "World's Best Low Cost Airline" in the annual World Airline Survey by *Skytrax* for 5 consecutive years from 2009 to 2013. Both these carriers, although different in profile, are proponents and exponents of the practice of disruptive innovation. Their philosophy has been the same: good economic sense; determination; being fearless of failure; excelling in marketing and branding; and always being on the watch out for new and untapped markets.

An effective approach to disruptive innovation is the exploitation of untapped markets. A good example is reflected in the post war Japanese motorcycle market where the four major manufacturers of the motorcycle market—Honda, Kawasaki, Suzuki and Yamaha had captured the market at the time the Japanese Diet passed an amendment to the Road Traffic Kaw in 1952 which allowed young drivers to ride motor cycles. Suzuki grabbed the opportunity and built the Diamond Free Bike—a low cc motor cycle for youngsters. Honda did the same with a similar product, both manufacturers selling their new product to a hitherto non-consumer market. This had an enormous impact on sales of the two companies as well as contributing vastly to the Japanese economy and boosting employment rates.[90]

The air transport industry's economic fortunes are cyclical and region-based. With the predicted growth in the global economy, particularly in the South East Asian region, a non-consumer market will emerge if it has not already done so. If anything, the story of air transport brings to bear two salient philosophical facts—If we do not adapt, we can become extinct; but changing the way we think, from protectionism to competition, is our biggest problem.

[90] Mezue et al. (2015), pp. 71–72. The authors state that Panasonic, Sharp and Sony in consumer electronics; Nissan and Toyota in cars; and Canon, Kyocera, and Ricoh in office equipment, all are examples of disruptive innovation.

References

Abeyratne R (2004) The decision in the *Ryanair* case – the low cost carrier phenomenon. Eur Transport Law XXXIX(5):585–601

Abeyratne R (2005) Emergent trends in aviation war risk insurance. Air Space Law XXX (2):117–129

Abeyratne RIR (1984) Competition and trade policies: their interaction. OECD, Paris, p 121

Abeyratne RIR (1993) The air traffic rights debate – a legal study. Ann Air Space Law XVIII(Part 1):3–44

Abeyratne RIR (1996) Legal and regulatory issues in international aviation. Transnational Publishers, New York, pp 271–313

Abeyratne RIR (1999a) Emergent commercial trends and aviation safety. Ashgate, Aldershot

Abeyratne RIR (1999b) Emissions trading as a market-based option in air transport – contractual issues. Environ Policy Law 29(5):226–235

Abeyratne RIR (1999c) The fuel tax and emissions trading as market-based options in air transport. Ann Air Space Law (Annales de Droit Arien et Spatial) XXIV:1–31

Abeyratne RIR (2001) Competition and liberalisation in air transport. World Compet 24 (4):607–637

Abeyratne RIR (2002a) International responsibility in preventing the spread of communicable diseases through air carriage – the SARS crisis. Transp Law J 30(1):53–80

Abeyratne RIR (2002b) The events of 11 September 2001 – ICAO's responses to the security and insurance crises. Air Space Law XXVII(6):406–420

Abeyratne RIR (2002c) Investing in insurance of air transport: some perspectives. J World Invest 3 (3):521–533

Abeyratne RIR (2002d) Attacks on America – privacy implications of heightened security measures in the United States, Europe, and Canada. J Air Law Commerce 67(1):83–115

Abeyratne RIR (2003) Sustainability of air carriers and assurance of services. J Air Law Commerce 68(1):3–37

Adkins B (1994) Air transport and E.C. competition law, European competition law monologues. Sweet and Maxwell, London, p 81

Aghion P, Bechtoldy S, Cassarz L, Herzx H (2014) The causal effects of competition on innovation: experimental evidence. Working Paper, February 20, 2014

Anthony S (2008) Why it's so hard to disrupt the airline industry. Harv Bus Rev, 8 March 2008

Benitah M (2001) The law of subsidies under the GATT/WTO system. Kluwer Law International, London, p 1

Boz E (2014) Why has global trade slowed down? World Economic Forum, Projections, https://agenda.weforum.org/2014/11/has-global-trade-slowed-down/

Constantinescu C, Mattoo A, Ruta M (2014) Slow trade. Finance Dev 51(4)

Davies G (2013) Why world trade growth has lost its mojo, September 29, http://blogs.ft.com/gavyndavies/2013/09/29/why-world-trade-growth-has-lost-its-mojo/

de Murias R (1989) The economic regulation of international air transport. McFarland, North Carolina, pp 52–72

Elgar E (2000) In: Kerr S (ed) Global emissions trading: key issues for industrialized countries. Northampton, MA

Elgar E (2001) International emission trading: from concept to reality. International Energy Agency, Paris

Ellenberg J (2014) How not to be wrong: the power of mathematical thinking. Penguin Press

Gander S, Helme N (1999) Emissions trading is an effective, proven policy tool for solving air pollution problems. ICAO J 54(7):12–14, 28–29

Ghemawat P (2011) World 3.0 – global prosperity and how to achieve it. Harvard Business Review Press, p 12

Goh J (1997) European air transport law and competition. Wiley, New York, p 15

Kupchan CA (2012) No one's world – the west, the rising rest and the coming global turn. Oxford University Press, New York, p 198
Lawton TC (2002) Cleared for take off, structure and strategy in the low fare airline business. Ashgate, Aldershot, p 1
Mezue BC, Christensen CM, van Bever D (2015) The power of market creation: how innovation can spur development. Foreign Aff 69:71–72
Rafiqul Islam M (1999) International trade law. Law Book Company, p 216
Schaal D (2015) The airline business is a lousy one, airline industry study shows, SKIFT: Megatrends Defining Travel in 2015
Schwab K (2015) Talent versus capital in the twenty-first century, Project Syndicate, http://www.project-syndicate.org/commentary/talent-capital-economic-growth-by-klaus-schwab-2015-05
Sharma R (2012) Breakout nations – in pursuit of the next economic miracles. Norton, p 242
Slaughter A-M (2004) A new world order. Princeton University Press, p 18
Spence M (2011) The next convergence: the future of economic growth in a multispeed world. Farrar, Straus and Giroux, p 36
Wassenbergh H (1996) De-regulation of competition in international air transport. Air Space Law XII:80
Weldon D (2015) Has the global economy slowed down? BBC News, 2 March 2015
Zakaria F (2011) The Post American World (Release 2.0). Norton, p 21

Chapter 6
Market Access and Subsidies in Air Transport: The US-UAE Debate and WTO

6.1 Introduction

The fact that market access in air transport does not come under the *Annex* to the General Agreement on Trade in Services of WTO, and the prevailing bilateral air services regime between States, which essentially precludes a totally liberalized air transport system, has led to protectionism for the past 70 years of regulated air transport. Although there are some open skies agreements between States, they have tended to ensure bilateral reciprocity for the most part. The subsidy issue between the United States carriers and the carriers of the United Arab Emirates (Emirates and Etihad) as well as Qatar Airways is the latest dispute where the United States' carriers claim that the governments of Qatar and the UAE have granted over $40 billion in concealed subsidies such as direct cash injections, interest free loans with no payment obligations, shareholder advances, significant related party transactions not at arm's-length, and subsidized infrastructure, as well as other unfair business practices, such as bans on unions. The US carriers claim that they are losing market share as a result of this exploitation which is causing the loss of hundreds of jobs. This chapter examines the issue in the context of market access and WTO rules.

6.1.1 The Debate

By letter dated 30 April 2015, 260 United States Congressmen wrote jointly to Secretary of State John Kerry and Secretary of Transportation Anthony Foxx urging them to "seek consultations with the governments of Qatar and the United Arab Emirates (UAE) in an effort to stem the tide of subsidized capacity that their state-owned airlines are deploying on international routes to the United States, in direct

R. Abeyratne, *Competition and Investment in Air Transport*,
DOI 10.1007/978-3-319-24372-6_6

contravention to the U.S. Open Skies policy".[1] The Congressmen went on to say: "Specifically, we have learned that over the past decade, the governments of Qatar and the UAE have granted over $40 billion in concealed subsidies, as defined by the World Trade Organization and U.S. trade law, have taken a wide variety of forms, such as direct cash injections, interest free loans with no payment obligations, shareholder advances, significant related party transactions not at arm's-length, and subsidized infrastructure, as well as other unfair business practices, such as bans on unions. The evidence is drawn almost entirely from financial statements, most of which have not been released in this country".

Curiously, the letter added: "*According to available research, each daily international round trip frequency lost/foregone by US airlines because of subsidized Gulf carrier competition results in a loss of hundreds of US jobs*".[2]

On the flip side of the coin, the airlines concerned take a different position. In its submission, responding to the claim of the US carriers, Etihad (The carrier of the United Arab Emirates) has claimed that what it received from the UAE government are loans and equity infusions which cannot be categorized as subsidies as claimed. Additionally, Etihad claims that in their argument, the three US carriers (American, Delta and United) do not claim that Etihad or any of the other carriers accused is engaging in a pricing war by lowering their prices. Also claimed by Etihad is that the US carriers' submission to Congress mentions the consumer only once (which means they are not concerned about the consumer's interest) and do not say that they have been hurt or have incurred any losses as a result of the UAE Carriers and Qatar operating air services in and out of the United States.

Maurice Flanagan, Emirates' founding managing director, said in a press interview in 2010:

> We have a stronger imperative to be profitable. I was given $10 million by Sheikh Mohammed to start the airline in 1985 and he said 'don't come back for any more, no subsidies of any kind whatsoever, no protectionism whatsoever.' In fact we have had $80 million in cash in kind since the start of the airline 25 years ago... Those were the rules of the game. You've got to be smart to succeed, smart even to survive.[3]

A compelling point made by Flanagan is that the story is quite the opposite of the subsidies claim of the West and that Emirates is not only not subsidised by the Dubai government but the airline in fact contributes significantly to the emirate's coffers, by paying $100 million in dividends to its corporate owner.

---

[1] http://pallone.house.gov/press-release/pallone-and-colleagues-lead-letter-opposition-abuse-open-skies-agreement.

[2] *Ibid.*

[3] IAG CEO and former head of British Airways, Willie Walsh, said he "doesn't see any evidence" that Gulf carriers receive huge state subsidies from their national governments. See *No evidence Gulf carriers get 'artificial support', says Willie Walsh* http://www.arabianbusiness.com/no-evidence-gulf-carriers-get-artificial-support-says-willie-walsh-564516.html#.VBUzWsWSxMg. Also, Interview: Maurice Flanagan, Emirates Airline, by Shane McGinley, July 27, 2010. See http://www.arabiansupplychain.com/article-4587-interview-maurice-flanagan-emirates-airline/2/.

"Every year we have paid more than a $100 million in dividend to the owner of the company," he says. "Emirates' latest annual report showed the Dubai government was… paid a dividend of AED956 million ($260 million), and AED2.91 billion ($792 million) in 2009". Flanagan mentioned the spat with Air Canada and its claim that Emirates was being subsidized and therefore could dump cheap capacity that would take away Air Canada's market share from Canadian destinations. "They are still there politically in the 1960s" Flanagan said: "I can't understand that. The market is there for double daily Toronto and double daily to Vancouver and certainly daily, going to double, to Calgary. We are [only] allowed three a week to Toronto and that is all protection to Air Canada [but] what good has that done to Air Canada? Look at the state they are in… I know of no other country that thinks in those terms. It is to protect Air Canada, I can't see any other reason. Just open the skies and let us in — that is the answer. It will be great benefit to the Canadian economy.[4]"

As part of its lobbying to the Canadian government, Emirates released a study in March 2010 concluding that Canada could reap economic benefits of around $466 million a year and create 2800 jobs if it was given the flying slots it is demanding. However, Air Canada has accused Emirates of wanting to "flood" the Canadian routes in order to divert passengers through Dubai.

James Hogan, Chief Executive of Etihad takes the same position for his airline. Abu Dhabi-based airline Etihad Airways of the United Arab Emirates applied for an open skies agreement with Canada. Etihad is keen to be able to expand its services out of Toronto. Hogan said at an interview:

> We've always been very clear that in business you have to respect your competitors. We're disappointed that these myths keep on being circulated… Our shareholder is the Abu Dhabi government, and as a shareholder they have invested in their airline via equity and start-up loans. Let's be clear. These are loans/seed capital. We have used that money to place fleet orders and build the infrastructure of the airline. And there is a repayment schedule for the loans. We have also raised $9 billion from 68 financial institutions, all without sovereign guarantees[5]

By 2010, the UAE, Dubai and Emirates Airline had secured over 60 open or highly liberal aviation agreements, following further liberalization success in Latin America, Africa and Europe. According to Tim Clarke, President of Emirates: "Open skies secured by the UAE and Dubai Civil Aviation Authorities with key economies now represent the majority of Emirates Airline's air services access worldwide, with an average of six new open or liberalized deals now being signed a year… The world, often led by emerging markets, is liberalizing faster than many believed possible. This is good news for consumers, traders, exporters and travelers generally".

When Emirates commenced its operations to Australia in 1997, the airline was viewed with trepidation and concern by QANTAS, as a threat to its market share.

---

[4] See Flanagan, *Ibid.*

[5] Walsh, *supra* note 3, *Ibid* (this chapter). For a general overview of civil aviation in the UAE, see Abeyratne (2008), pp. 3–15.

This concern was shared by the Australian authorities. However, attitudes quickly changed, and this concern was obviated when they realized the added economic benefit quickly enjoyed by the places Emirates flew to. Currently, Emirates flies to five cities across Australia, and offers onward flights to New Zealand from several of these. QANTAS and Emirates are now partners.

The Economist states that the legacy carriers of the West have not been destitute of State protection themselves, adding that the super connectors from the middle east have a competitive advantage in their locations, massive airports that could connect millions of passengers between east and west, efficient jets, young and keen staff, and low costs on their long haul to long haul routes.[6]

Although international economic law is seemingly a patchwork of provisions emerging from various agreements and amendments,[7] a deeper examination reveals that State subsidies are governed under a central theme which discourages unfair trade practices. Subsidies, which are government grants or bounties, are an integral part of international trade and entitle a government, by a selective process, to assist trading services and entities to the betterment of society.

Whoever is right in this debate is not the issue. The trouble with air transport is that, while on the one hand it is a product, on the other hand regulations pertaining to this product may constrain its availability to the consumer by depriving him of the various choices of air travel he might have under a liberalized system. In other words, State policy and the protection of national interests take precedence over the interest of the user of air transport. The aviation industry offers only one product to the ultimate consumer and that is the air transport product.

"Connectivity" which is the most compelling need in aviation, and embodied in the Chicago Convention as *inter alia* "meeting the needs of the people of the world for efficient and economical air transport" is stultified by interests of commercial and national policy. The International Chamber of Commerce (ICC), in a policy statement has expressed the view that the efficiency of air transport would be enhanced by creating more open markets and more flexibility with regard to foreign ownership. Given air transport's capability to facilitate economic activity, its liberalization would enable the sectors that make use of it to become non efficient. ICC was in favour of a freer exchange of air services throughout the world and is convinced that it is time to move beyond the existing bilateral system, toward a genuine multilateral liberalization of air transport.[8]

The US Congressmen's letter makes one wonder whether the fate of air transport lies in internal job creation and not in connecting the world which is the main intent and purpose of the Chicago Convention.

---

[6] Super Connecting the World – The Advance of Emirates, Etihad and Qatar, latterly Joined by Turkish Airlines looks set to continue, *The Economist* April 25, 2015, 60–61 at 61.

[7] Benitah (2001), p. 1.

[8] ICC Policy Statement: The need for greater liberalization of international air transport Prepared by the Committee on Air Transport, *Document n° 304-2/23 Rev.3*, 1 December 2005 at 5.

At the heart of the debate is ICAO which is charged with the task of *fostering the development of air transport*, whatever that means. For 70 years, ICAO has maundered in circumlocution and dithered on a way forward with just empty words on "promoting and encouraging liberalization". The 6th ICAO Air Transport Conference in March 2013 recognized ICAO as the only forum for initiating global solutions for the development of a sustainable air transport system.

Yet, one notices a glaring lack of foresight and responsibility when this statement is tied to the opening speech at the conference of the President of ICAO's Council who said: "Given aviation's historic role in supporting improved social development and economic prosperity, the potential benefits of growth are enormous over the coming decades. There is a serious chance these benefits won't fully materialize, however, unless we come up with practical and concrete recommendations for adapting the global regulatory framework to the realities of the 21st century".[9]

Those "practical and concrete recommendations for adapting the global regulatory framework" are yet to attain fruition. In the meanwhile, ICAO stands impotent in trying to give effect to its constitution—the Chicago Convention[10]—which, in its preamble states *inter alia* that international air transport services may be established on the basis of equality of opportunity and operated soundly and economically.[11]

A way forward that has been suggested as a practical bypass to ICAO's feckless insouciance in this regard—that is to inquire into the possibility of including market access under the umbrella of the World Trade Organization (WTO)[12] which

[9] ICAO News Release, ONCE-A-DECADE ICAO AIR TRANSPORT CONFERENCE CONVENES, MONTRÉAL, 18 March 2013.

[10] *Supra*, Chap. 1, note 5.

[11] See generally, Abeyratne (2013a), pp. 9–29. Also see Abeyratne (2013b); and Abeyratne (2014), pp. 8–11 and 23.

[12] The home page of the World Trade Organization (hereinafter referred to as WTO) succinctly summarises its nature and scope: There are a number of ways of looking at the World Trade Organization. It is an organization for trade opening. It is a forum for governments to negotiate trade agreements. It is a place for them to settle trade disputes. It operates a system of trade rules. Essentially, the WTO is a place where member governments try to sort out the trade problems they face with each other. See https://www.wto.org/english/thewto_e/whatis_e/who_we_are_e.htm. The World Trade Organization (WTO-OMC) came into being on 1 January 1995 and along with it entered into force the General Agreement on Trade in Services (GATS). The GATS Annex on Air Transport Services applies trade rules and principles such as most-favoured nation (MFN) treatment and national treatment to three specific so-called "soft" rights, namely, aircraft repair and maintenance, selling and marketing of air transport, and computer reservation system services. It excludes from the application of the GATS "services directly related to the exercise of traffic rights". Pursuant to an earlier ministerial decision, the WTO-OMC launched in 2000 the first review of the operation of this Annex with a view to considering possible extension of its coverage in this sector. During the review, which continued into 2002, there was some support to extend the Annex to include some additional "soft" rights (for example, ground handling) as well as some aspects of "hard rights" (for example, air cargo, non-scheduled and multi-modal transport). However, there is no global consensus at this stage on whether or how this would be achieved. See also, https://www.wto.org/english/docs_e/legal_e/26-gats_01_e.htm.

superseded the General Agreement on Tariffs and Trade (GATT).[13] This article looks at the principles that would apply to subsidies for airlines if the subject of market access in air transport were to be remanded to WTO. The contentious issue of subsidies in aviation is not new to WTO. A few years ago, the WTO tribunal adjudicated on the issue of subsidies given to Boeing (by the United States) and Airbus (by European governments).[14]

6.1.2 *The United States and Europe*

6.1.2.1 United States

The *Sherman Antitrust Act* of 1890[15] of the United States (US) prohibits "unreasonable restraint" in trade and unlawful monopolization, particularly between states within the US and with foreign nations. The *Foreign Trade Antitrust Improvements Act* of 1982 extended the spirit and scope of the *Sherman Antitrust Act* by making the former applicable to foreign nations but the validity of this legislation as being applicable extra territorially has been judicially questioned in the US. Nonetheless, it has been applied to an instance which occurred in US territory.[16]

The *Air Transportation Competition Act* (*Deregulation Act*) of 1979 deregulates air transport in the US and reduces statutory barriers that stand against any carrier from entering new international markets, while giving flexibility to carriers to alter their fares without let or hindrance from the regulator. Of note is a feature in the Act that grants authority to the government to take action against a foreign government which practices anti competitive conduct and discriminatory practices against American carriers and their right to operate air services.

It is apparent that none of these statutory provisions are applicable to UAE carriers operating air services into and out of the US in the face of the Open Skies Agreement between the US and UAE. The UAE carriers cannot be curbing

---

[13] Former specialized agency of the United Nations. It was established in 1948 as an interim measure pending the creation of the International Trade Organization. However, plans for the latter were abandoned and GATT continued to exist until the end of 1995. Members of GATT were pledged to work together to reduce tariffs and other barriers to international trade and to eliminate discriminatory treatment in international commerce. The eighth, or Uruguay round, of GATT negotiations, which began in 1986 with 15 negotiating groups, was long stalemated by the issue of agricultural subsidies maintained by the European Community. The agreement that resulted (1994) from the Uruguay round led to the creation (1995) of the more powerful WTO. However, the GATT framework remained in place for a 12-month transition period.

[14] In May 2003 The United States called for WTO to appoint a panel of judges to go into the claim of the United States that European governments were illegally subsidizing Airbus Industrie, the Toulouse based manufacturing consortium which builds the Airbus range of aircraft. See Abeyratne (2005), pp. 379–395.

[15] !5. U.S. SS 1.

[16] *United States* v. *Sital Sales Corp*, 274 U.S. 268 at 275–76.

competition which the US carriers are free to engage in untrammelled. On the contrary, the *Deregulation Act* promotes and encourages lower fares and higher productivity. The Act succeeded in facilitating US carriers injection of more capacity on routes.[17]

The *Air Transportation Safety and System Stabilization Act*, signed into law by President George W. Bush as a response to the events of 11 September 2001, established *inter alia* the Air Transportation Stabilization Board which was authorized to issue US $10 billion in subsidies based on credit instruments to airlines, mainly in the form of loan guarantees. There were conditions imposed in relation to these subsidies such as requiring the bankrupt free status of applicant carriers and their submission of credible forecasts and evaluations for growth. The Act also admitted of Presidential powers to compensate by executive order carriers incurring losses as a result of slow traffic growth in the aftermath of 9/11, such compensation not to exceed US $5 billion.

6.1.2.2 Europe

European Union policy generally forbids a member State of the Union to subsidize airlines with State aid, on the basis that such subsidies would be anti competitive. However, if an airline faces difficulties, there is some justification to invoke an exception that would admit of subsidies. Against this backdrop, the overall rationale that prohibits anti competiveness would apply as a general rule that effectively precludes subsidies in the Union. It must be noted that as a counter measure, the liberalization of air transport by the European Council in 1993 and the total removal of restrictions in 1997 acts as giving an impetus within the Union for airlines to have a fair and equal opportunity to compete with each other.

In the *Sabena* case[18] The European Commission allowed Belgium to grant aid to the restructuring of *Sabena*, on the condition that there would be no involvement of the Belgian government in the management of the airline, and that there would be no further injections of cash aid to the airline. There was a similar granting of aid to *Iberia* by the Spanish authorities in 1992 and 1995. The Commission in 1994 granted *TAP Air Portugal* tax exemptions and permission to increase capital.[19] Some conditions were imposed such as the mandatory applicability of the relief to certain areas in Portugal.

Another significant milestone in the EU State aid equation is the approval given by the European Commission to France to implement a compensation scheme to

---

[17] A.E. Kahn, *Airline Deregulation, The Concise Encyclopaedia of Economics*, Library of Economics and Liberty. See http://www.econlib.org/library/Enc1/AirlineDeregulation.html.

[18] [1991] OJ L 300/48, 31.10.2001.

[19] Commission Decision 94/698/EC, 6 July 1994.

buttress *Air France* when it had incurred significant losses after the closure of certain areas of air space as a result of the events of 9/11.[20] The application of stringent security measures in response to the 9/11 attacks was also considered an exigency that constrained normal air traffic services. However, The State aid given under the circumstances was eventually withdrawn in view of the incompatibility with compensation rules within the Union on the ground that additional costs incurred as a result of circumstances arising out of the 9/11 attacks and heightened security measures could not be considered as exceptional.[21]

The European Commission brought an action in 2003 against Greece for granting illegal aid to *Olympic Airways* and for not recovering such aid from the airline.[22] In 2011 a similar measure was imposed by the Commission on the Hungarian authorities for giving subsidies to *Malev*—the Hungarian national carrier—where the former made capital injections to the carrier and gave loans to shareholders of the airline company. Such impositions by the European Commission have led European governments to a dilemma of sustaining their carriers amidst strong competition on the one hand and abiding by anti-competitive rules of the European Union on the other.

Article 85.1. a) of the Treaty of Rome[23] pronounces that any agreement which distorts competition is void. The owner of Charleroi Airport [24]—the Walloon Region—signed an agreement with *Ryanair* on 6 November 2001, giving the airline a reduction of approximately 50 % of the landing charge at the airport which had ordinarily been fixed by the Walloon Government for all carriers serving the airport. The common rated charges base had been fixed by a decree issued in 1998 by the

---

[20] See Report *COM (2001) 574 Final*. The Report contained a discussion on the consequences of the 9/11 attacks on the air transport industry.

[21] The decision was taken pursuant to the *Treaty on the Functioning of the European Union (TFEU)* Article 108.1 and 108.2 of which provide that The Commission is authorized to, in cooperation with Member States, keep under constant review all systems of aid existing in those States. It can propose to the latter any appropriate measures required by the progressive development or by the functioning of the internal market. If, after giving notice to the parties concerned to submit their comments, the Commission finds that aid granted by a State or through State resources is not compatible with the internal market, or that such aid is being misused, it is obligated to decide that the State concerned shall abolish or alter such aid within a period of time to be determined by the Commission.

[22] *Commission* v. *Hellenic Republic*, Case C-415/03.

[23] The Treaty of Rome, officially the Treaty establishing the European Economic Community (TEEC), is an international agreement that led to the founding of the European Economic Community (EEC) on 1 January 1958. It was signed on 25 March 1957 by Belgium, France, Italy, Luxembourg, the Netherlands and West Germany.

[24] Brussels South Charleroi Airport (BSCA) was at the time a public sector company contracted by the Walloon Region, which has managed the airport under a 50-year concession agreement.

Government and was applicable to airport taxes that included landing, passenger and parking fees.[25] It was claimed that the reduction granted was discretionary on the part of the Walloon Minister of Transport, by means of a private contract by-passing the requisite statutory process. The deviation from established practice was, it was alleged, not in keeping with the required objectivity as the charges were calculated on the basis of each embarking passenger, instead of the usual tonnage weight of the aircraft. Furthermore, as an integral part of the agreement between the Walloon Region and *Ryanair*, the former had undertaken to compensate the latter for any loss of profit arising from a change in the level of airport taxes that might occur during the years 2001–2005 as a result of an internal decision taken by the Region. Under the agreement, the airport was authorized to keep 65 % of the fees collected by the Walloon Region from airlines serving the airport. The Belgian authorities had allowed this retention on the basis that the airport would undergo expenditures in welcoming, embarkation and disembarkation of passengers.

Part of the European Commission's decision on the *Ryanair* case was that Belgium had unlawfully given aid to Ryanair in contradiction of Article 88 (3) of the Treaty of Rome. This notwithstanding, the Commission, in appreciation of the possible contribution such aid can make in the launching of new air transport services along with the sustainable development of a regional airport, found it acceptable that a portion of such aid could be considered compatible with the objectives of the Common Market provided *inter alia* such aid was publicized; the total aid was calculated; compensation guarantees for financial losses to *Ryanair* were void; and contributions paid to *Ryanair* beyond a 5-year period were repaid by *Ryanair*.[26]

6.2 Market Access and WTO

6.2.1 ICAO Perspectives

The issue as to whether market access in air transport should be included in the General Agreement of Trade in Services (GATS) Annex of the WTO has been

[25] Decree of the Walloon Government of 16 July 1998 laying down the fees to be levied for the use of airports within the Walloon Region, Moniteur Belge of 15 September 1998. See 2004/393/EC, Commission Decision of 12 February 2004, reproduced in the *Official Journal of the European Union*, L 137/1 (30.4.2004) at p. 2.

[26] For an indepth discussion of the Ryanair case see Abeyratne (2004), p. 585.

discussed in ICAO for quite some time.[27] The GATS Agreement,[28] in Part 1 Article 1.2 defines a trade in service *inter alia* as the supply of a service from the territory of one Member into the territory of any other Member. Part II Article 2 of the Agreement provides that, with respect to any measure covered by the Agreement, each Member is required to accord immediately and unconditionally to services and service suppliers of any other Member, treatment no less favourable than that it accords to like services and service suppliers of any other country. This provision is diametrically opposed to the principle enunciated in a typical bilateral air services agreement that gives a grantor State the prerogative to prescribe capacity, routes and frequencies to a foreign carrier without such a "most favoured nations treatment" requirement.

At the ICAO Air Transport Colloquium held in 1992 several views were presented, mostly favouring protectionism and the then prevailing bilateral air services negotiations system which still applies. The current system is based on national treatment, where States could apply different conditions to carriers operating air services into their territories based on capacity and demand for travel at prices that are deemed acceptable to the State concerned and the travelling public. Corollaries to the current system of bilateral air transport agreement are traditionally seeped in norms requiring capacity to be primarily provided for traffic to and from the two States parties to the agreement, on the basis of "fair and equal opportunity". Under this system, airlines are usually expected to be substantially owned and effectively controlled by a minimum percentage of the Contracting party. This affords a State the opportunity to protect the interest of its own carriers.

At the Fifth ICAO Worldwide Air Transport Conference held in 2002, a *Declaration of Global Principles for the Liberalization of International Air Transport* was adopted (by ovation) where one of the critical and thought provoking provisions is found in clause 4.4, which provides that each State will determine its own path and own pace of change in international air transport regulation, in a flexible way and using bilateral, sub-regional, regional, plurilateral or global avenues according to circumstances. Given that the overall approach of the Fifth Worldwide Air Transport Conference was "how to liberalize" (as against "whether to liberalize" which was the preoccupation of the earlier Air Transport Conference in 1994), this provision seems to say that the issue of how to liberalize is very much left to the

[27] GATS seeks to liberalize trade regarding the provision of services. It deals with, and provides for two types of obligations—general and specific. The former relates to obligations of States and the latter to commitments made by States. Also, The GATS Agreement identifies and makes provision for specific services through its various annexes. The Air Transport Annex of GATS currently includes aircraft repair and maintenance services; the selling and marketing of air services; and computer reservations systems. Notwithstanding the profile of the Annex, a certain ambivalence pervades the interpretation of inclusionary elements in the Annex and their application. This is evident in the fact that only a few WTO members have committed to the three activities. Only three aspects of air transport are included under the GATS umbrella: computer reservation systems; sale of air transport and repair of aircraft and components.

[28] https://www.wto.org/english/docs_e/legal_e/26-gats_01_e.htm.

States themselves, to be done at their own pace. If this be the purport and intent of this provision, it does not say anything much except to endorse what States had been doing prior to the Conference. One wonders in this context what the intent of the Contracting States of ICAO was, in adopting this platitude.

A repetition of the intent of this clause is found in the provision immediately proceeding clause, where States are requested that they should, to the extent feasible, liberalize international air transport market access, ensure air carrier access to international capital and air carrier freedom to conduct commercial activities—again a truism and fact of economic reality that had been happening in the aviation community before the Conference. Regarding cargo services liberalization, the same type of "endorsement" is seen in the statement that States should give consideration to liberalizing the regulatory treatment of international air cargo services on an accelerated basis, provided that clear responsibility and control of regulatory safety and security oversight is maintained.

At the 6th ICAO Worldwide Air Transport Conference (ATConf/6) held in 2013, the Conference decided, as further action to "meet the needs of the people of the world for safe, regular, economical and efficient air transport" (as prescribed in the Chicago Convention to continue to assist States in their liberalization efforts by enhancing the "market place" facility offered to States; continue to update the ICAO Template Air Services Agreements (TASAs) to keep pace with regulatory evolution; undertake and promote the development of additional training courses, regional seminars or similar activities for the benefit of States, in accordance with available resources; continue to monitor regulatory developments, conduct studies on major issues of global importance, and provide policy guidance and assistance to States; and continue to develop relevant databases such as the Database of the World's Air Services Agreements as well as case studies of liberalization experiences.

The issue for international air transport is "what is the central market place?" The immediate answer which comes to mind, if air transport were to be treated as a trade, is the WTO.[29] One of the most contentious issues in the world of commercial air transport today is the question as to whether the industry should embrace the trade in services regime of WTO in preference to the currently restrictive system of the bilateral air transport agreement which entitles States to refuse permission to air carriers who apply to operate commercial air services to and from their territories. The protectionism of States based on market share that their national carriers arrogate to themselves, and the overall misconception that inclusion of market access will lead to total liberalization seems to be what is keeping States from treading this path. There are other misconceptions, as the discussion below reflects.

[29] For general reference, see Abeyratne (1998), pp. 39–56. Also Abeyratne (2003), pp. 633–667.

6.2.2 Some Misconceptions

One reason for the lack of clarity that permeates the issue of market access in air transport services, particularly with regard to its relation to WTO, is the proliferation of misconceptions that tend to obfuscate public policy debates on the possible role of GATS in liberalizing air transport within the parameters of the Air Transport Annex. This ambivalence may partially be due to a lack of familiarity within the aviation community of the complex web of rules governing trade practices within the purview of WTO regulations and also due to the pervading protectionism that has been exhibited by nations irrespective of their resource bases when it came to market access.

The first misconception that has to be addressed is that which pertains to the general belief that inclusion of market access related issues of air transport in the GATS would necessarily lead to liberalization. If air transport services are covered by the Annex, it would indeed remain a voluntary domestic decision of the State concerned and coverage under the GATS Annex would not imply deregulation or any attendant obligation to revise and modify existing regulatory regimes. A commitment to provide air transport services under the GATS Annex would essentially retain for the State concerned its pristine right to enforce regulations in force that bring to bear obligations governing safety, environmental protection and security. Furthermore, committing to a GATS governed market access system would not impel a State to alter or in any manner derogate from entrenched principles regarding foreign ownership of airlines.

Another misconception associated with the GATS system is that it is a rigid, inflexible mechanism that would stultify individual regulatory reform within a State or inhibit a State from initiating its own legislation in trade related issues such as safety, security and environmental protection. The GATS offers the air transport industry, through their States' mechanisms, the flexibility to make choices based on material interests.[30] The GATS would offer member States the right to select opportunities and times to make sector specific market access decisions and national treatment commitments. The GATS admits of progressive liberalization, in accordance with differing levels of development of services. Given the exponential growth of the air transport industry, this system would effectively facilitate consistency between national initiatives toward progress and predatory practices associated with excessive competition.

It must be noted that the most fundamental purpose of GATS is to provide for the liberalization of trade pertaining to the provision of services. GATS seeks to establish a multilateral framework of principles and rules for trade in services

[30] See generally, Market Access: Unfinished Business, Post Uruguay Round Inventory and Issues, World Trade Organization, Special Studies 6:2001 at pp. 99–105.

with a view to expansion of such trade under conditions of *transparency*,[31] national treatment[32] and *progressive liberalization*.[33] The fundamental principle of GATT is its Most Favoured Nation (MFN) Treatment clause[34] whereby each party to the agreement accords immediately and unconditionally to services and service providers of any other party, treatment no less favourable than that it accords to like services and service providers of any other country. These provisions reflect the basic philosophy of GATS and play a vital role in affecting the decision of the international community on whether or not air transport services should be brought under its purview. Other features of GATS which have attracted discussion in relation to air services are provisions relating to increasing participation of developing countries within GATS[35] and dispute settlement.[36]

For the present, the overall purpose of including air transport services in GATS seems to be to apply the broad principles of market access and the MFN philosophy to the selling or marketing of air transport services. The purview of GATS in controlling air transport services would therefore be considered only in situations where air traffic rights are exercised multilaterally or plurilaterally. GATS would not apply in instances where States elect to use Article 6 of the Chicago Convention which governs all bilateral air services agreements and requires that the permission of a grantor State is necessary for a commercial air transport enterprise to operate air services in to or out of a State. In any event, the Annex on Air Transport Services to GATS does not reflect confidence in itself by providing in Article 6 of the Annex that the operation of the Annex shall be reviewed periodically or at least every 5 years.

---

[31] Article III of GATS requires each party to publish promptly all relevant laws, regulations, administrative guidelines and all other decisions, rulings or measures of general application, by the time of their entry into force.

[32] GATT's national treatment philosophy provides foreign services and services suppliers with treatment no less favourable than that accorded to a country's own services and service suppliers.

[33] Since GATS is an Annex to the GATT agreement it should be noted that the provisions of GATS are governed by those of GATT and that both documents incorporate the same basic principles.

[34] Under the WTO agreements, countries cannot normally discriminate between their trading partners. If one state grants another a special favour, it would have to do the same for all other WTO members.

[35] Article II of GATS. Article XVI extends the MFN principle to market access.

[36] Article XXIII on dispute settlement is considered to be well balanced and equitable and provides that if any Party should consider that another Party fails to carry out its obligations or commitments under the agreement, it may make written representations or proposals to the other Party or Parties concerned and the latter should give sympathetic consideration to the representations or proposals so made. If no satisfactory settlement could be arrived at, the GATT agreement provides for a formal dispute settlement procedure in Articles XXII and XXIII.

6.3 Subsidies and WTO

The term "subsidy" is not defined precisely in economic terms although the *Oxford English Dictionary* defines the word as "a sum of money granted from public funds to help an industry or business keep the price of a commodity or service low". In broad terms therefore, a subsidy can be considered indirect protectionism. Under the *Agreement on Subsidies and Countervailing Measures* (SCM Agreement),[37] a subsidy is recognized as a financial contribution by a government which confers a benefit.[38] A financial contribution is either money or anything else of value provided to a manufacturer or exporter (which could be construed as an international air service originating in a country) at a cost less than would have been charged in a commercial transaction. This could include indirect support from a government.[39]

Sykes is of the view:

> . . .depending on the nature of the subsidy program, subsidies may lead recipients to reduce prices and expand output. Such behaviour by a subsidized firm will attract customers away from unsubsidized firms. It follows immediately that subsidies can be used to protect domestic firms from foreign competition. A subsidy to domestic firms that compete with imported goods or services can induce them to expand their share of the market relative to the share of imports, in much the same way as a tariff or a quota.[40]

WTO is driven by The SCM Agreement which is an extension to the *Agreement on Interpretation and Application of Articles VI, XVI and XXIII* (12 April 1979) which was negotiated in the Tokyo Round which took place in the seventies and was the first major attempt to tackle trade barriers that do not take the form of tariffs, and to improve the system. The latter agreement set out the basic principle—that subsidies are used by governments to promote important objectives of national policy, but that subsidies may have harmful effects on trade and production. The emphasis of the Agreement was on the effects of subsidies and that these effects are to be assessed in giving due account to the internal economic situation of the signatories concerned as well as to the state of international economic and monetary relations. The aim was to ensure that the use of subsidies does not adversely affect or prejudice the interests of any signatory to the Agreement, and that countervailing measures do not unjustifiably impede international trade, and that relief is made available to producers adversely affected by the use of subsidies within an agreed international framework of rights and obligations.

One commentator is of the view:

> . . .government interventions are subject to strict regulations under the Agreement on Subsidies and Countervailing Measures (SCM AGREEMENT) and other relevant rules under the World Trade Organization (WTO). As argued by the Japanese government, the

[37] https://www.wto.org/english/tratop_e/scm_e/subs_e.htm.

[38] Article 1.1.

[39] Cunningham (1999), p. 6.

[40] Sykes (2003), p. 8. http://www.law.uchicago.edu/Lawecon/index.html.

> provision of subsidies is lawful as a general rule and is not limited to those to the MRJ (Mitsubishi Regional Jet) project. But Article 5 of the SCM AGREEMENT prohibits WTO members from causing "adverse effects" to other members by providing subsidies. In particular, "serious prejudice" referred to in Article 5(c) (and defined in Article 6) may arise, for instance, in the case where a like product of another member economy is displaced by the subsidized product or if the subsidized product is deemed to have caused a significant decrease in the price and/or sales volume of a like product of another member country.[41]

In practicality, one can see the validity of this statement in the Brazil-United states dispute before the Dispute Settlement Board of the WTO. Brazil averred that the US provided prohibited and actionable subsidies to US producers, users and/or exporters of upland cotton, as well as legislation, regulations, statutory instruments and amendments thereto providing such subsidies (including export credits), grants, and any other assistance to the US producers, users and exporters of upland cotton. Brazil contended that these measures were inconsistent with the obligations of the United States under the SCM Agreement. The WTO Panel appointed by the Board found that the United States acted inconsistently with its obligations under Articles 5(c) and 6.3(c) of the SCM Agreement in that the effect of marketing loan and counter-cyclical payments provided to US upland cotton producers.[42]

Assuming the UAE subsidizes its two airlines and Qatar does likewise, one could inquire whether serious prejudice is caused by the American carriers replacement (if such replacement is established) by the three airlines operations into the United States. Emirates Airline's President Tim Clarke said in an interview in 2013 that UAE has an open skies agreement with the U.S. which allows the airline to carry passengers on a fifth-freedom basis from the West Coast and central points in the U.S. to points in Asia.[43] The United States government says of its open skies agreements:

> Open Skies agreements have vastly expanded international passenger and cargo flights to and from the United States, promoting increased travel and trade, enhancing productivity, and spurring high-quality job opportunities and economic growth. Open Skies agreements do this by eliminating government interference in the commercial decisions of air carriers about routes, capacity, and pricing, freeing carriers to provide more affordable, convenient, and efficient air service for consumers.
>
> America's Open Skies policy has gone hand-in-hand with airline globalization. By allowing air carriers unlimited market access to our partners' markets and the right to fly to all intermediate and beyond points, Open Skies agreements provide maximum operational flexibility for airline alliances.[44]

---

[41] Tsuyoshi Kawase, State Support for Aircraft Development and the WTO Rules on Subsidies: On the occasion of the rollout of the first Mitsubishi Regional Jet, http://www.rieti.go.jp/en/columns/a01_0409.html.

[42] https://www.wto.org/english/tratop_e/dispu_e/cases_e/ds267_e.htm#top. Also, https://www.organicconsumers.org/old_articles/corp/cotton042904.php On 16 October 2014, Brazil and the United States notified the DSB that, in accordance with Article 3.6 of the DSU, they had concluded a Memorandum of Understanding, and agreed that this dispute was terminated.

[43] Flottau (2013). *See* http://aviationweek.com/awin/uae-carrier-plans-maximize-open-skies-agreements.

[44] http://www.state.gov/e/eb/tra/ata/.

The open skies agreement signed by and between the aeronautical authorities of the United States and The UAE in 2002[45] stipulates, in Article 11 that Each Party will allow a fair and equal opportunity for the designated airlines of both Parties to compete in providing the international air transportation governed by the Agreement and that each Party will allow each designated airline to determine the frequency and capacity of the international air transportation it offers based upon commercial considerations in the marketplace. Consistent with this right, neither Party is allowed to unilaterally limit the volume of traffic, frequency or regularity of service, or the aircraft type or types operated by the designated airlines of the other Party, except as may be required for customs, technical, operational, or environmental reasons under uniform conditions consistent with Article 15 of the Chicago Convention.[46] There is no restriction by way of unfair competition caused by State aid in the Agreement. This is where The WTO principles would take over if market access is brought into its regime.

Pursuant to part III of the *SCM Agreement*, subsidies that are specific may be challenged if they cause economic harm by way of adverse effects which can result in injury to the domestic industry of another Member; nullification or impairment of benefits accruing directly or indirectly to other Members; or serious prejudice to the interests of another Member. The traffic light approach to subsidies adopted in the SCM Agreement is clear. Subsidies falling under the "green light" are allowed. The "yellow light" subsidies are not permissible if they adversely affect or otherwise hinder free trade. Subsidies falling under the "red light" category are not tolerated. The economic harm categories laid out in the first part of this paragraph are yellow light subsidies. Suitable analogies to yellow light subsidies are the cotton subsidies of the US that formed part of the discussion above and any subsidies that might be given by a State to an airline that would be calculated to undermine fair competition in the provision of air transport services.

Even if the $42 billion subsidy alleged to have been given by the UAE to its carriers is proven, it would be tenuous to bring it under the WTO rules as discussed above, should market access be brought within the purview of the GATS Annex. The SCM Agreement specifies that a subsidy is deemed to exist if there is a financial contribution by a government or any public body within the territory of a Member, i.e. where: a government practice involves a direct transfer of funds (e.g. grants, loans, and equity infusion), potential direct transfers of funds or liabilities (e.g. loan guarantees); government revenue that is otherwise due is foregone or not collected (e.g. fiscal incentives such as tax credits); a government provides goods or services other than general infrastructure, or purchases goods; a

---

[45] http://www.state.gov/documents/organization/125743.pdf.

[46] Article 15 of the Chicago Convention states *inter alia* that every airport in a contracting State which is open to public use by its national aircraft shall likewise ... be open under uniform conditions to the aircraft of all the other contracting States. The like uniform conditions shall apply to the use, by aircraft of every contracting State, of all air navigation facilities, including radio and meteorological services, which may be provided for public use for the safety and expedition of air navigation.

government makes payments to a funding mechanism, or entrusts or directs a private body to carry out one or more of functions which would normally be vested in the government and the practice, in no real sense, differs from practices normally followed by governments; or there is any form of income or price support in the sense of Article XVI of GATT 1994[47]; *and a benefit is thereby conferred* (my emphasis).

Even if all other conditions of the SCM as reflected above are met, The US carriers would have a difficult time proving that a benefit has been conferred on the UAE carriers through subsidies granted to them. Under the US/UAE Open Skies Agreement there is no distortion of competition nor is there a specific benefit in operating air services as allowed in the Agreement. There is no prohibition of subsidies nor any mention thereof in the Agreement. On the contrary, the services offered by the UAE carriers, and Qatar Airways as well as Turkish Airlines (which has also been identified as a threat to US carriers), only stimulate competition.

The air transport industry stands at the crossroads of two major influences—globalization and the information revolution—which have revolutionized the trading world by driving competition. The fact that the UAE carriers (as well as Qatar Airways and Turkish Airlines) have the geographic advantage of being in the centre of the long air routes of the world should not be taken as an argument supporting competitive distortion or disadvantage to other carriers which would in turn unduly curb the operations of the four carriers. The Air Transport Association (IATA)[48] has recognized that US $4.2 trillion is needed over the next two decades to properly offer connectivity to a travelling public who are creating an exponential demand for air transport.[49]

IATA's Director General Tony Tyler has said that what is needed is smart regulation from governments around the world in order to maximize the economic benefits of connectivity—jobs and growth. Tyler says that unfortunately, high taxation and poorly designed regulation in many jurisdictions make it difficult for airlines to develop connectivity. One of the problems identified by Tyler is that in addition to cost issues, airlines also face a hyper-fragmented industry structure owing to government policies that discourage cross-border consolidation. He suggests fresh thinking on all accounts.[50]

One need not say more.

---

[47] Article XVI provides that any subsidies affecting exports to and imports from Members should be notified in writing. Members should recognise the deleterious impact of subsidies and avoid their general and specific use.

[48] The International Air Transport Association is a trade association of the world's airlines. These 250 airlines, primarily major carriers, carry approximately 84 % of total available air traffic.

[49] Schaal (2015). See http://skift.com/2013/07/01/the-airline-business-is-a-terrible-one-says-leading-airline-industry-group/.

[50] *Ibid.*

References

Abeyratne R (1998) Competition rules governing market access in air transport: a comparative study. World Compet 21(4):39–56
Abeyratne R (2003) Trade in air transport – have we lost our way? J World Trade 47(3):633–667
Abeyratne R (2004) The decision in the "Ryanair" case – the low cost carrier phenomenon. Eur Transport Law: J Laws Econ 39:585
Abeyratne R (2005) The airbus Boeing subsidies issue – some preliminary legal issues. Air Space Law XXX/6:379–395
Abeyratne R (2008) Civil aviation in the United Arab Emirates – some legal and commercial perspectives. J Air Law Commerce 73(1):3–15
Abeyratne R (2013a) Reinventing ICAO's role in air transport – a compelling need? Issues Aviat Law Policy 13(1):9–29
Abeyratne R (2013b) Regulation of air transport: the slumbering sentinels. Springer, Heidelberg
Abeyratne R (2014) Should ICAO have a role as an economic regulator of air transport? Air Space Lawyer 27(1):8–11 and 23
Benitah M (2001) The law of subsidies under the GATT/WTO system. Kluwer Law International, p 1
Cunningham RO (1999) Subsidies to large civil aircraft production: new WTO subsidy rules and dispute settlement mechanism alter dynamics of U.S.–E.U. dispute. Air Space Law 6
Flottau J (2013) UAE carrier plans to maximize open skies agreements. Emirates wants to expand its network by adding transpacific routes. Aviation Week & Space Technology, June 10
Schaal D (2015) The airline business is a lousy one, airline industry study shows. SKIFT: Megatrends Defining Travel in 2015
Sykes AO (2003) The economics of WTO rules on subsidies and countervailing measures. The Law School, The University of Chicago. John M. Olin Law & Economics Working Paper No. 186 (2d series) May, p 8

Chapter 7
Achieving Competitive Advantage Through Connectivity and Innovation: An Application in Airline Hubbing

7.1 Introduction

Latest technological advances, coupled with changing needs and the growing demands of customers, have forced organizations to develop and produce better services. Globalization,[1] which essentially removed trade barriers between nations, bringing to bear the blatant reality that increasing global connectivity, together with integration and interdependence in the economic, social, technological, cultural, political and ecological spheres, have removed all trade barriers, making a world without boundaries, has not worked exactly that way in the air transport industry. Protectionism of market share by States for their national carriers is still alive and well in spite of the fact that globalization brings with it the inevitable corollary of the contemporaneous advancement of technology and growing trends toward liberalism in international trade. The information and telecommunications revolution, which really kick started in the 1980s, dramatically lowered the costs of doing business across national borders. The giant strides made by information technology, which took its incipient steps in the 1990s, together with paradigm shifts in trade practices such as outsourcing and off-shoring, have ensured the opening of a world which no longer sees boundaries that inhibit global trade and information exchange.

The strongest thrust of globalization in the business world is its ability to generate competition within and between nations to offer the best goods and services at the lowest prices. However, the woes of the air transport industry have always centered on the claim that aviation is effectively precluded from entering the free market due to government meddling. This seems to go contrary to the Keynesian view of the advantages of some degree of government control of the free market economy. Harsh restrictions on ownership and control of airlines and market access are stringent tools of government control.

[1] Encyclopedia Britannica defines globalization as the process by which the experience of everyday life is becoming standardized around the world.

R. Abeyratne, *Competition and Investment in Air Transport*,
DOI 10.1007/978-3-319-24372-6_7

The above notwithstanding, air transport confers indispensable and significant benefits to the global economy. In 2012, the industry contributed US $2.4 trillion to the global GDP (3.4 %). In addition, about US $606 billion has been accrued as direct benefits to the world economy from air transport (i.e. employment and economic activity generated by the air transport industry) and US $697 billion are indirect benefits (generated by employment and economic activity of suppliers of the air transport industry). A key benefactor of these contributions are major hubs such as Singapore and Dubai. In Dubai, for instance, aviation generates about 28 % of the city's GDP.[2]

Aviation also plays a key role in enabling the economic growth of countries which rely on connectivity to the world. A report[3] issued in 2014 by OECD[4] reflects that although some countries have liberalized their air transport policies, there are some challengers of a regulatory nature that still remain to stultify air transport and effectively preclude it from being a true globally connected industry that would be governed by a seamless and integrated international regulatory framework. OECD notes the existence of a trend towards a significantly more heterogeneous series of economic regulatory frameworks reflecting national priorities and bilateral agreements between individual States, and concludes that this limits innovation, investment and expansion in aviation markets. Quite correctly, OECD opines in its report that the ultimate aim of liberalisation of air transport is to obviate these limitations with a view to benefitting the consumer of the air transport product and promoting his interests and welfare.[5]

To stay afloat, an air carrier has to constantly innovate and re-strategise. This is mostly because competition between carriers is intense and often disruptive, calling for strategic responses to technological advances, customer demands and globalized competition. Although some airlines find a good niche and apply a robust plan to develop it and consolidate themselves, others have to seek partnerships to establish new markets. Others who are failing need to refocus and renew their strategy. All these scenarios bring to bear a compelling need to constantly re-strategise corporate strategy owing to the rapidly changing exigencies of the air transport business.[6] There have been many definitions[7] of competitive advantage, and, broadly speaking, a competitive advantage confers economic value to a competitor over others in the business. One of the ways airlines have gained a

---

[2] Arvis and Shepherd (2011), p. 15.

[3] *Liberalisation of Air Transport Summary: Policy Insights and Recommendations*, International Transport Forum, at 4.

[4] Organisation for Economic Co-operation and Development (OECD), established in 1961, promotes policies that are calculated to improve the economic and social well-being of people around the world. The OECD provides a forum in which governments can work together to share experiences and seek solutions to common problems. The Organization works with governments to understand what drives economic, social and environmental change.

[5] *Liberalisation of Air Transport Summary, supra* note 3 (this chapter), *loq.cit.*

[6] See A Palette of Plans, *The Economist*, May 30th 2015, at 66.

[7] Rumelt (2003), p. 1.

competitive advantage is through a sustained and well established organizational culture[8] established mainly as a result of strong leadership exercised with consistency and the use of clear messages and signals that convey corporate priorities, beliefs and values to both the corporate staff and the customer. Corporate culture,[9] particularly in airlines, is a valuable tool that helps in establishing and maintaining a brand that in turn leads to organizational growth. In the airline industry, the product of which makes the consumer psychologically vulnerable, clear establishment of ethical values that assure the customer of safety, security and comfort is critical.

There is also a school of thought that the empowerment of employees in a corporate body helps establish competitive advantage on the basis that employee empowerment facilitates the creation of an integrated quality environment, where superior products and services become practical. This is accomplished by sharing information, creating autonomy, and establishing self-directed teams.[10] Yet others claim that companies do not gain competitive advantage due to their ignorance of their own competencies and the knowledge and expertise of their staff which can be used to establish the company firmly in a market niche.[11]

The business factors discussed above form the collective basis for innovation that would circumvent regulatory challenges faced by airlines who seek a competitive advantage over their competitors while promoting connectivity that would meet the needs of the travelling public more efficiently. There are some airlines that have used disruptive innovation[12] through their hub operations to gain a competitive advantage over their competitors and take command of an existing market. This article discusses competitive advantage within the concept of airline hub operations and innovation.

---

[8] Organizational culture is defined as a "dynamic phenomenon that surrounds us at all times, being constantly enacted and created by our interactions with others and shaped by leadership behaviour, and a set of structures, routines, rules, and norms that guide and constrain behaviour". See Schein (2004), p. 21.

[9] See Goksoy et al. (2013), p. 304.

[10] Kahreh et al. (2011), p. 27.

[11] King et al. (2001), p. 95.

[12] "Disruptive innovation" is a term introduced by Clayton Christensen—a Harvard professor—and is essentially descriptive of a process by which a product or service establishes itself with a value network as a new entrant at the bottom of a market and then relentlessly moves up market, eventually displacing established competitors. See http://www.claytonchristensen.com/key-concepts/.

7.2 The Need for Connectivity in Air Transport

Air traffic growth, in which the protagonists were once North America and Europe, is now led by the Middle East, Asia-Pacific region and Latin America which have experienced strong growth over recent years. IATA,[13] in 2013 reported that whilst revenue passenger kilometres (RPKs)[14] in North America and Europe have grown at a rate of 2.2 % and 4.0 % respectively, the Middle East, Asia-Pacific region and Latin America have shown growth rates of above 6 % per annum, specifically 11.9 %, 7.2 % and 6.5 %. Even in Africa, growth in air transport has been significant showing an increase of 5.1 % in RPKs since 2012. Most of the growth is be attributed to economic growth as well as to regulatory changes which allow for greater market access.

ICAO identifies connectivity as "the ability of a network to move a passenger from one point to another with the lowest possible number of connections and without an increase in fare, focusing on, from a commercial perspective, minimum connecting times with maximum facilitation ultimately resulting in benefits to air transport users".[15] At its 6th Air Transport Conference held in September 2012, ICAO noted that there were four key players in air transport: Member States, aircraft operators, airport operators, and air transport users, all with different priorities, and that they had one thing in common—the pursuit of connectivity in air transport.[16] This definition notwithstanding, it must be noted that connectivity should not only be addressed in the context of the passenger but in the perspective of the carriage by air of cargo (freight) as well. The World Bank, in its *Air Connectivity Index* of 2011 defines connectivity as "the importance of a country as a node within the global air transport system". A country is considered to be better connected the stronger is the overall "pull" it exerts on the rest of the network.[17] Air connectivity is measured using a variety of measures at various levels of granularity. These measures—including total passenger movements, airfares, the number of direct destinations, and travel time—can serve as stand-alone proxies or may be combined to create a measure capturing different features of the air-transport market. The *Index*, which is based on the premise that liberalization leads to better connectivity, also makes the point that it is extremely important to account for hub and spokes interactions when measuring connectivity in the air transport network. This view accords closely with thinking in the industry.[18] It

[13] The International Air Transport Association (IATA) is the trade association for the world's airlines, representing some 250 airlines or 84 % of total air traffic. We support many areas of aviation activity and help formulate industry policy on critical aviation issues.

[14] RPKs are a common industry measure of air traffic which is calculated by multiplying the number of revenue-paying passengers aboard the vehicle by the distance travelled.

[15] Enhancement of Air Transport Connectivity and Facilitation, *ATConf/6, WP-20*, 19/12/12, at 2.

[16] *Id*, 1–2.

[17] Arvis and Shepherd (2011), p. 3.

[18] *Id*. 10.

suggests that countries with strong connectivity will specialize in industries such as perishable goods and networked components, which are intensive in their use of high speed transport services. Furthermore, it is expected that air transport connectivity will influence exporters' choice of transport mode as well.[19]

In the ultimate analysis, one should ask the inevitable question: what is aviation's strategic direction? As far back as 1944 The Chicago Convention,[20] intent on "creating and preserving friendship and understanding among the nations and people of the world" declared in its Article 44 that the world should strive, through ICAO, to meet the needs of the peoples of the world for safe regular, efficient and economical air transport. One would not imagine, for a moment, that there is written anywhere in a global document that aviation's strategic direction is to make as much money as possible to the exclusion of others or to give priority to the interests of States or the air transport industry or any other business. Creating and preserving "friendship and understanding among people" can only be achieved through optimum connectivity.[21]

There is no room for doubt that connectivity in air transport is critical to increasing g a country's economic growth potential, by attracting business investment opportunities and human capital. Another important factor is the boost given to tourism by connectivity, which in turn contributes significantly to a country's economic growth.[22] According to Lipman, connectivity is key to the growth of tourism and therefore growth of the world economy. Lipman claims that "travelism's (travel and tourism coined) influence – positive and negative – has a significant place in global growth transformation; representing directly and indirectly some 5 %–10 % of world socio-economic activity, with Africa's 9 % GDP and 8 million jobs contribution forecast to grow annually at close to 5 %. With new pan-African/regional collaboration and China-led bilateral and multilateral funding, we can expect to see major road, rail and air links across the continent to spur the movement of the increasing middle class and an ultimate market of 1 billion consumers".[23]

---

[19] *Id.* 37.

[20] *Supra*, Chap. 1, note 5.

[21] Article 6 is arguably the most contentious in the Chicago Convention. In brief, it contends the following: all scheduled international commercial air transport services are prohibited except to the extent they are permitted; all bilateral and open skies agreements are reciprocal and subject to the nuances of aeropolitics and protectionism and arbitrary demarcations of market share; therefore airlines do not have freedom to access of markets; there are rigid and archaic ownership and control regulations governing so called "national carriers; In many instances, this effectively precludes direct foreign investment in airlines; all the above unduly prevent connectivity, which is the meaning and purpose of meeting the needs of the people of the world for regular, efficient and economical air transport.

[22] Morphet and Bottini (2012), p. 11.

[23] Geoffrey Lipman, *How The Travel and Tourism Industry Can Influence Sustainable Growth*, http://www.eturbonews.com/59877/how-travel-and-tourism-industry-can-influence-sustainable-growth.

Air connectivity has four main drivers which are geographic locations; airport infrastructure and facilities; business models of airlines and economic regulation in States concerned. These four aspects are also the main determinants of hubbing.

7.3 Hubbing

The term "hub" in an air transport sense has been ambivalent at best. Originally, the term was used to identify any large airport or a large base of a commercial airline. However, with the advent of deregulation of air transport (both in the United States and elsewhere) and more recently, open skies agreements between States and groups of States, the word "hub" has taken a more precise operational meaning. According to one commentator, a hub is "an integrated interchange point where one or more specific airlines concentrate traffic and operate waves of flights".[24]

Airlines use the hub and spokes system of operating air services (as graphically demonstrated by the hub and spokes design of a bicycle wheel) as the antithesis of operating air services on a point to point basis. The Hub-and-spokes system came about as a result of the US Airline Deregulation Act of 1978, prior to which airlines operated the origin-destination, point-to-point routing which was often not cost efficient. The concept of the Hub-and-spokes system uses one centrifugal airport (or more than one airport in the case of multiple hubs) that can distribute spokes-flights to smaller airports.[25] An example is the carriage of passengers from New York to Mumbai via Dubai (hub) where Mumbai is the spoke. With this system airlines are able to serve more markets with existing resources, particularly crew and fleet. The hub and spokes concept is also strategic and invariably profitable and obviates airlines' obligations to operate unprofitable point to point flights. In addition, airlines could optimize results of hubbing through strategic alliances with other airlines, consolidating resources and air traffic rights.

Arguably, the most spectacular strategic airline alliance so far is the "Star" Alliance, which was launched in 1997 by Lufthansa, SAS, United Airlines, Thai Airways International and Air Canada. Brazilian carrier Varig joined later. The underlying philosophy of the airline alliances, typified by a "Star" Alliance, is not so much an emphasis on the more effective use of resources such as labour, capital and national resources (which are inevitably important factors) but rather an overall

---

[24] Danesi (2006), p. 55.

[25] One commentator puts it aptly: "The change in market entry and exit rules altered the way airlines schedule service. Prior to deregulation, it was quite common for passengers to change airlines (interlining) in order to reach a destination because the origin city and destination city were not served by the same airline. With deregulation, each airline was able to expand to serve more cities, and it accomplished this by offering a network pattern called *hub and spokes*. In this pattern, flights came in from several cities to a centrally located hub airport. At the hub, the passengers switched planes on the same airline and continued their trip out to the final destination". See Grandeau (1995), p. 11.

reliance on the strategy of location, where the sharing of locations represented by the various airlines have enabled them to produce their goods and services in a consistent manner, thus achieving the status equivalent to a cartel, while still retaining their individual identities.

Airlines have developed both a corporate strategy and a competition strategy to cope with competition. Both these strategies are becoming increasingly complementary rather than being mutually exclusive, which they were at the inception of airline competition 50 years ago. As airlines began to compete with each other across the borders, they acquired the ability to locate themselves overseas—creating a compelling need for commercial airlines to be fully acquainted with locational strategy and competitive advantages of various locations.

The world's biggest international air hubs are London Heathrow, Paris Charles de Gaulle, Dubai International, Hong Kong and Amsterdam Schiphol. An example of a combination of strategic alliances and hubbing is American Airlines, Delta Air Lines, and United Airlines which carry onward traffic through European hubs in cooperation with European airlines with whom they have formed strategic alliances. These airlines have structured their businesses in such a way that a substantial portion of their international traffic passes through Europe; specifically, through the hub airports of their respective European partners.

Airlines operate through hubs by manipulating their slots (the day and time of arrival of an aircraft and departure therefrom) schedules. For instance, banking of slots became popular with deregulation where airlines used hubs at peak times for traffic to optimise flights and compete with point to point carriers. However, since banking of schedules meant that aircraft and staff would idle at de-banked hours this process proved uneconomical and airlines began to de-peak their schedules at hubs. One commentator states that "de-peaking schedules was one way airlines responded to the high costs they faced throughout the 2000s. By reducing the maximum number of aircraft that depart and arrive at an airport within a period of time, gates, equipment, and personnel can be used more efficiently. Inactivity is reduced as aircraft are constantly being serviced. American Airlines was the first U.S. airline to implement de-peaking across parts of its system in order to control for costs. In 2002, American de-peaked its hubs in Chicago O'Hare (ORD) and Dallas/Fort Worth (DFW), responding to the economic downturn post-9/11".[26]

A regression analysis conducted on the co-relation between hub concentration and costs has revealed that an increase in hub concentration yields a statistically significant increase in cost per available seat mile (CASM), a statistically significant decrease in cost per passenger seat mile (CPSM), and a statistically significant decrease in operating expenses. The CASM increase is attributed to the fact that when opportunity is afforded to airlines to operate more flights in a concentrated hub scenario they have to pay more in airport fees to help the hub airports finance large capital improvement projects, which in turn increase an airline's total costs. Where the CPSM decrease is concerned, this is due to the application of economies

[26] Katz (2012), p. 4.

of scale, where fixed costs can be spread through more passengers from whom revenue is derived.[27]

As for airports, which act as hubs, they provide the community with better choices in the nature of a wider network of routes and regular frequency, paving the way for increased connectivity. Hubs create a win-win situation for both connecting passengers with increased choices, reducing travel time and costs by offering more direct and more frequent links, particularly in long haul travel. The larger the size of the hub the better facilities and choices offered and one large hub generates more connectivity than the sum of two hubs of half the size.[28] The salient features that determine the success of a hub are: central geographical location vis-à-vis the most important traffic flows and feeder airports; peak hour capacity to facilitate an efficient wave-system structure of the hub airline; a strong hub carrier being part of a global airline alliance; availability of traffic rights (market access); Short Minimum Connecting Time One Terminal concept; competitive visit costs; good landside accessibility; available options for future growth; and airport amenities.[29]

In certain instances where competition is intense and the operations of an airline are significant on long haul operations, there could be a need for multiple hub operations. Airlines could utilize and manipulate their frequencies in operations on long thin routes that require multiple frequencies owing to excessive demand. Called the "frequency game", this traffic pattern enables the airline to enjoy high yield traffic in the hubs.[30] One of the most advantageous features of hubbing is that it increases traffic density and therefore long haul flights through hubs become attractive to the consumer, for the reason that increased traffic density, particularly achieved with the use of large aircraft, decreases the seat mile cost for the airline and therefore reduces the price of a seat. Just one city added to a network increases the provision of that new city to every other city in the network. This allows the airline better capacity utilization.

A competitive advantage in hubbing is that whether in a single hub or multiple hubs, passenger choice is a factor that would determine the stronger competitor. Flight choice of a passenger is driven by airfare, convenience of schedule, the airline brand and the promises the airline makes in terms on time performance and on board facilities as well as aircraft type. Leisure passengers are more price conscious than business passengers while the business passengers prefer a convenient flight schedule and extra comforts both on board and on the ground in terms of the transit lounge.[31] In order to achieve this significant business advantage carriers that use hubs have to increase their efficiencies and lower their labour costs.[32]

---

[27] Short (2013), p. 40.

[28] Burghouwt (2013), p. 10.

[29] *Id.* 12.

[30] Goedeking (2010), p. 17.

[31] See Kim (2005), p. 23.

[32] Cook (2008), p. 57.

7.4 Innovation

Innovation in terms of economic growth of any industry can be categorized into three forms: *sustaining innovation*—which replaces old products with new, keeping an established market growing further and offering an improved product (examples are new and improved models of television sets, Iphones); *efficiency innovation*—products which help industries produce more quantities with less resource expense (a wide spread retailer selling the same product as a designer brand store for a lesser price); and *market creating innovation*—products which are offered for a more attractive price than identical products offered elsewhere, which creates an entirely new market (targeting the non-consumer, for example the introduction of the Harry Potter series in books and films which targeted the young and early teens).[33] There is a fourth category—*disruptive innovation*[34]—a business concept which is an innovation that helps create a new market and value network that disrupts the existing market.

There are four legacy carriers that have, through hubbing, used both market creating innovation and disruptive innovation. They are Emirates Airlines, Etihad Airways, Qatar Airways and Turkish Airlines. With a lower cost base and the geographic advantage of their hubs these four airlines have used the buzzword "connectivity" to displace, to a large extent, the attraction offered particularly by carriers of the western world on long thin routes. They have also tapped the non-consumer market in their disruptive innovation. The non-consumer market can be seen in countries such as China and India, the two most populous countries in the world, where in the former the growing middle class (of ages 20–50) from households with an annual income of between $10,000 and $60,000 U.S. dollars are starting to travel. This category of consumer which the world saw emerge between the last 10–15 years, is estimated that it's more than 300 million—already larger than the entire population of the United States. About 25 % of the population of China is middle class which comprises about 50 % of the urban population.[35] India, which has a burgeoning middle class of more than 500 million, is the world's fastest

---

[33] For a discussion on these three innovations see Mezue et al. (2015), pp. 69–70.

[34] For a definition see *supra* note 12 (this chapter).

[35] In a 2012 CNN interview the person interviewed said: "The Chinese are shopping a lot more. Retail is booming like a wildfire in China. There are a lot more consumers and they are demanding a lot more services. A lot of Chinese, especially younger consumers, are really into the luxury brands. They associate Western luxury brands with quality of life and sophistication. They want gyms, health care clubs and definitely travel. They want to see the world. The restaurant business is doing very well. The younger generation—people under 30—they are consuming like crazy. They save zero. They spend all of their salary on a Louis Vuitton purse. A lot of them stay with their parents so they don't have housing expenses. But once they get married, then they start to save". See Tamy Luhby, *China's growing middle class*, http://money.cnn.com/2012/04/25/news/economy/china-middle-class/index.htm.

growing economy, according to Fortune Magazine which reported this fact in February of 2015.[36]

The three carriers in the Persian Gulf (Emirates, Etihad and Qatar) are carrying a major component of the traffic between the Middle East and Southeast Asia with the assistance of their hubbing activities in their backyards. They successfully injected increased capacity between Thailand and their States by 80 % in 2014, while achieving a capacity increase in Europe by just 4 %, resonating the fact that the hubs in the Middle East are being made increasingly popular by these carriers. As against the success of these carriers, Thai Airways could only fill just 5 % of the seats between Thailand and the United Arab Emirates, and the three Gulf carriers have acquired the other 95 % of market share.[37] It is also noteworthy that the three carriers in the middle east had also disrupted the market of giant Singapore Airlines between the Gulf and the far East by shrinking Singapore Airlines' share of traffic to and from the Middle East from 55 % in 2009 to 15 % in 2013, where the three carriers accounted for two-thirds of the total available seats on flights from Singapore to their three hubs.[38]

All three UAE carriers have successfully attracted the emerging market that is flowing from the West to the Indian Sub Continent (ISC) as well as China and the Far East, by using a robust and competitive marketing strategy which has innovatively created strong brand awareness worldwide. They sponsor sports clubs and events in both the UAE and around the world, having established sponsorship as a critical stimulus to and essential feature of their marketing strategy. Their new orders of Airbus A380 aircraft which can carry more than 550 persons at a time, and strategically excellent geographic location, enable them to continue to grow and serve their competitors' main hubs. As one commentator said in 2010: "Emirates are a great case to watch. The leadership of a strategy that exploited a technology, a large new aircraft, the A330 that could operate from smaller regional airports and the use of Dubai with its time zone advantage in offering a good stop-off and transshipment point on the way to the East, has been a masterful exploitation of the white space in the existing airlines businesses".[39] Added to this is the fact that all three Gulf carriers achieve their innovation through the recruitment of talented and skilled staff.

The three carriers of the Gulf are classic examples of disruptive innovation where, in an era of high competition and volatility in the business world, they are not only selling their airlines (which other competitors do as well) but also marketing an image that their competitors do not have. Inasmuch as the largest hotel operator Airbnb has no hotels and the biggest taxi operator Uber, has no taxis,

---

[36] http://fortune.com/2015/02/27/its-official-sort-of-india-to-overtake-china-as-fastest-growing-major-economy/.

[37] Kindergan (2015).

[38] *Ibid.*

[39] https://leaderswedeserve.wordpress.com/2010/06/21/emirates-airline-the-secret-story-of-a-successful-company/.

AliBaba no inventory, and Facebook makes billions with no content of its own, the three carriers are slightly different in that although they do have airlines (and excellent ones at that) their main strength is in their hubs and what they offer therein. The three airlines compete heavily with each other for their hub-traffic, which is indicative that they are on their own in terms of competition.

Market creating innovation and disruptive innovation—two related but not identical concepts which reflect the attributes of successful hubbing by airlines—can be identified with *The Game Theory* which is a philosophy drawn on the discipline of applied mathematics, that could be applied to air transport. The Game Theory—a quantum theory on anticipatory intelligence—is about maximising returns based on the strategic decisions to be made by contestants at economics, trade or politics. In air transport, the theory would help analyze interactions of carriers and strategies between them, thus enabling the airlines competing with each other to study strategic interactions between them. The outcome is a formal modelling approach to economic situations in which decision makers interact with other decision makers. The three UAE carriers have used this concept to successfully take advantage of the open skies agreement between the UAE and the United States, which gives them an insight into their competitors strategy, and formulate their own strategy accordingly with regard to hubbing. As a result, one would find it very difficult to prove anti competitive conduct on the part of these carriers.

There is however, a caveat. The one stumbling block to competition in air transport, which has bred controversy and contention, is the absence of an accurate perspective and understanding of the term "fair competition". No one has defined the term and the closest one has got to is in the Chicago Convention which states in its Preamble that air transport services may be established on the basis of equality of opportunity and operated soundly and economically. There has been no explanation as to what "equality of opportunity" means. Some have posited that it is equality of opportunity to compete with each other (in which case hubbing would become contentious) while others have said it should be interpreted as equality of opportunity to operate air services (in which case too hubbing would be brought into question, although the UAE carriers have contended that any carrier can use their airports as a hub).

Competition is associated with the rights of the competitor as well as the consumer, whereby the former is precluded from exercising dominance over others in the overall competitive environment, by acting in a manner calculated to eliminate, restrict or deter competition.[40] Such action is termed "predation" and is particularly associated in the field of air transport with pricing.[41] Unfair competition, which is believed to result in one competitor being a dishonest or fraudulent rival in trade and commerce, usually refers to misrepresentation of one's product through deceptive packaging, labeling, or other characteristics such as pricing of

[40] Hanlon (1994), p. 91.

[41] Roos and Sneek (1997), p. 154.

goods and services offered. In a broad sense, unfair competition refers to any activity that unfairly creates an advantage. Competition law *de lege ferenda*, which is law as it should be if the rules were changed to accord with good policy, would dictate that in a state of "perfect competition", there would be a benchmark for evaluating performance in actual markets. Economists contend that under conditions of perfect competition, goods and services would be produced as efficiently as possible—that is, at the lowest possible price and cost—and consumers would get the maximum amount of the goods and services they need.[42]

Unfortunately, the economic variants in air transport are far too complex to be treated on the basis of "perfect competition". Therefore, at best, the air transport industry can look to a "workable competition" regime, such as the one introduced by American economist John M. Clark in 1940, through which he recognized that in most industries the number of business firms is not so great as to preclude an individual firm from having some power to influence market prices and conditions.[43] In addition, participants rarely have complete knowledge of market conditions. Clark theorized, however, that departures from the ideal of perfect competition are often not great enough to warrant government intervention into the market (through antitrust action or direct regulation) in order to improve the situation. He further asserted that competition may be workable in the sense that the results achieved could be approximately comparable to the outcome of the theoretical ideal of perfect competition in application. The main difficulty emerging from the workable competition concept is its vagueness, in that no precise criteria have been developed to determine when workable competition actually exists. This is certainly true of the air transport industry and the esoteric competitive advantage bestowed to an air carrier through of the practice of hubbing.

References

Adkins B (1994) Air transport and E.C. competition law. Sweet & Maxwell, London, pp 89–90

Arvis J-F, Shepherd B (2011) The air connectivity index – measuring integration in the global air transport network. Policy Research Working Paper 5722, p 3

Burghouwt G (2013) Airport capacity expansion strategies in the era of airline multi-hub networks. OECD Discussion Paper No. 2013-5, February 2013, p 10

Cook GN (2008) Airline networks: a comparison of hub-and-spokes and point-to-point systems. J Aviat/Aerosp Educ Res 17(2):57

Danesi A (2006) Measuring airline hub timetable co-ordination and connectivity: definition of a new index and application to a sample of European hubs. Eur Transport\Trasporti Europein 34:55

[42] Adkins (1994), pp. 89–90. Gill and Bates (1949), p. 64. Uittenbogaart (1997), p. 222. Dempsey (1992), p. 206.

[43] This model was the precursor to the modern oligopoly theories that are now used to represent the structure in airline markets.

Dempsey PS (1992) Airline deregulation in the United States: competition, concentration, and market Darwinism. Ann Air Space Law/Annales de Droit Arien et Spatial XVII(Part I):199–243
Gill FW, Bates GL (1949) Airline competition. Harvard University, Boston, p 64
Goedeking P (2010) Networks in aviation. Springer, Heidelberg, p 17
Goksoy A, Vayvay O, Ergeneli N (2013) Gaining competitive advantage through innovation strategies: an application in warehouse management processes. Am J Bus Manag 2(4):304
Grandeau SC (1995) The processes of airline operational control. Master's Thesis submitted to the Massachusetts Institute of Technology, February, p 11
Hanlon P (1994) Discriminatory fares: identifying predatory behavior. J Air Transport Manag 1 (2):89–102
Kahreh MS, Ahmadi H, Hashemi A (2011) Achieving competitive advantage through empowering employees: an empirical study. Far East J Psychol Bus 3(2):27
Katz DS (2012) Revenue and operational impacts of depeaking flights at hub airports. PhD Thesis, Georgia Institute of Technology, December, p 4
Kim J (2005) An agent-based model for airline evolution, competition and airport congestion. PhD thesis, May, Falls Church, Virginia, p 23
Kindergan A (2015) REVISITING: the rise of the gulf carriers. The Financialist, January 2, 2015, https://www.thefinancialist.com/the-rise-of-the-gulf-carriers/
King AW, Fowler SW, Zeithaml CP (2001) Managing organizational competencies to competitive advantage: the middle management edge. Acad Manag Exec 15(2):95
Mezue BC, Christensen CM, van Bever D (2015) The power of market creation: how innovation can spur development. Foreign Aff 94(1):69–70
Morphet H, Bottini C (2012) Air connectivity: why it matters and how to support growth. PwC| Connectivity and growth, p 11
Roos HB, Sneek NW (1997) Some remarks on predatory pricing and monopolistic competition in air transport. Air Space Law XXII(3):154–157
Rumelt RP (2003) What in the world is competitive advantage? Policy Working Paper 2013-105, August 5, p 1
Schein EM (2004) Organizational culture and leadership, 3rd edn. Jossy-Bass, p 21
Short DM (2013) An empirical analysis of airline network structure: the effect of hub concentration on airline operating costs. Thesis, Duke University, Durham, North Carolina, p 40
Uittenbogaart P (1997) Airline competition on the route between Amsterdam and London. J Air Transport Manag 3(4):217–225

Chapter 8
Competition and Open Skies

8.1 Introduction

When one addresses the issue of open skies and the competitive aspects of the practice, the agreement between the United States and the European Union presents itself as the best example. Before the US/EU Open Skies Agreement was finalized and adopted carriers of both parties were circumscribed by stringent market restrictions that distorted competition between carriers of the United States and European countries. The European carriers themselves had to contend with the fact that they could fly directly to the United States only from their own country, which limited competition on transatlantic routes and within the European Union. Furthermore, EU carriers from Open Skies countries could not merge with those from non-Open Skies countries without losing US traffic rights, which thwarts integration and liberalisation of the internal EU market, further limiting competition. Mergers of US and EU carriers were prohibited altogether, which put many of the benefits of globalisation out of reach.

Against this scenario, The Brattle Report, released by the Brattle Group in 2007 identified some qualitative advantages that the open skies agreement between the US and EU brought. Among these was the fact that the now known business practice of creative destruction would come into play, enabling more efficient companies to replace less efficient firms. The Report said:

In a liberalised market, more efficient airlines would replace less efficient ones, or less efficient airlines would adopt the practices of more efficient ones, leading to significant cost savings and an increase in industry efficiency. This substitution would occur through two mechanisms: industry restructuring (e.g., mergers, acquisitions, joint ventures), and increased competition (e.g., a carrier from one EU country could establish a transatlantic hub in another EU country).

The Report also submitted that efficiency gains by carriers operating under the open skies scenario would be able to exploit size-related economies, leading to further efficiency gains. An example cited is the instance of a merger or deep

R. Abeyratne, *Competition and Investment in Air Transport*,
DOI 10.1007/978-3-319-24372-6_8

alliance which could enable two carriers to spread certain fixed costs over more passengers (scale economy). The carriers could also reconfigure their combined network to bring about more connectivity by moving more flights to certain hub airports. They could also achieve higher utilisation by combining traffic to raise load factors (density economy).

Another advantage identified by the report was that, by forging closer Integration among airlines, airlines could attain synergies in pricing, particularly by facilitating deeper forms of integration between US and EU carriers, where liberalisation would admit of improved price coordination on transatlantic interline routes (i.e., routes that require passengers to fly on two or more airlines to reach their destination). Without this new flexibility brought to bear by the open skies agreement, each airline would be prompted to set the fare for its leg of the flight without considering what the pricing effect would be and demand would be for the other legs. If the same carriers were to coordinate with each other, each will have an incentive to set lower fares so as to increase combined profits—a process that is now known to be the elimination of double marginalisation.

Arguably the most compelling—in support of the fundamental objective of competition, which is the welfare of the consumer—was that cost savings from the above advantages would be passed through to consumers, particularly in the long run where he could benefit from lower prices and enhanced connectivity. There was also the possibility that these advantages would increase demand, both among the high paying passengers and those who sought economy. The liberalisation of aviation trade and investment likely would lead to significant cross-border flows of capital as airlines engage in consolidation and deeper integration and establish new operations in markets that are opened or made more accessible by liberalisation. Cross-border investment would in turn play a major role in driving many of the benefits described above. The other qualitative advantage the Brattle Report identified concerned the cross border proliferation of labour that a liberalized open skies agreement would bring. The Report said that what was known as labour substitution, i.e.: substitution of less expensive foreign workers for more expensive domestic workers either directly or indirectly.

One notes that the American version of "open skies" is conducive to open competition as it comprises, *inter alia*, the basic elements of open entry to all routes; unrestricted capacity and frequency on all routes; unrestricted route, traffic rights, double disapproval pricing in third and fourth freedom markets in intra EC markets; liberal charter arrangements; liberal cargo regimes; open code sharing opportunities; and explicit commitment on non-discriminatory operation of access to computer reservation systems. Although in the nineties Europe thought this concept of open skies and its elements would "endanger the whole process of deregulation of Europe's civil aviation market", the overall approach of the EU to the negotiations with the US has been to accept the philosophy of the US on open skies. However, the scales would tip in favour of the US carriers under this philosophy, as they would be able to consolidate fifth freedom rights through a network of routes whereas the European carriers would not have a comparable variety of points to exercise fifth freedom rights beyond the US.

Another inhibitor for European carriers could be the anti-trust exigencies that arose under US law which may put European carriers at a substantial disadvantage in the US market. For this and the more significant distinction between US ownership and control restrictions (which are placed at 25 % for foreign nationals) and leverage given to foreign nationals (49 %) in the EU, it would be structurally, economically and legally more advantageous to US carriers to have an open skies agreement with Europe while their European counterparts may not be as well placed under such an agreement. Therefore an open skies agreement with EU might just help the American carriers in improving their revenues which have shown a $9.1 billion loss in 2004, followed by a $10 billion loss in 2005 and a projected $6.5 billion loss in 2006. In contrast, the European carriers have shown profits in the triennium, with $0.6 billion in 2004, $1.3 billion in 2005 and a projected $0.6 billion in 2006.

One of the commercial considerations with regard to achieving enhanced competition through an US/EU open skies agreement was the extent to which the EU would have the authority to negotiate all aspects of the DoT standard open skies agreement, which include slots. Another would be whether individual EU member States could still, after the EU signs an open skies agreement with the US, negotiate a bilateral air services agreement with the US From the EU perspective, where the Council of Ministers has given a mandate for the Commission to negotiate an agreement with the US, there is no express prohibition so long as there is recognition of the EU as one single area and the Community clause is signed. The consequence of the European Court of Justice decision, particularly from a competition angle, was that the core element of the bilateral air services agreement, which is market access involving the award of air traffic rights was untouched by the Court except in instances where an EU member would, in its agreement with the US, explicitly preclude another EU member from operating air services from that member's territory. In other words, Belgium would not be permitted to agree that Air France would not or could not operate services between Brussels and New York. This prohibition is entrenched in the Treaty of Rome which forms the substance of legislative legitimacy of the EU and incorporates the right of equal national treatment for all EU member States. Therefore, if one EU member State were to preclude the right of another member State's airline from having the right to operate air services to the United States from the territory of the first EU State it would tantamount to discrimination by the first State against the second State.

Also, the Court decided that certain specific provisions and areas covered in the questioned bilateral agreements between individual EU members and the US were contrary to EU law since they encroached upon internal EU regulations pertaining to non EU nationals. These laws concern:

a) provisions pertaining to the allocation of airport slots;
b) provisions governing pricing, or fares and rates of intra European air services;
c) agreements on computer reservation systems insofar as they appear as provisions of the open skies agreements in question; and

d) provisions which reserved the right to grant permission under the open skies agreements only to airlines substantially owned and effectively controlled by nationals of the EU member States that is party to a particular agreement.

Yet another issue to be considered was the position of the United States and the EU on the issue of public subsidies. This extends both to carriers as well as manufacturers. On the one hand, one recalls the Anglo-French Concorde which sustained its services through subsidies by the British and French governments. On the other, both parties are in dispute over subsidies purportedly given to Boeing and Airbus by the US and EU respectively. Further impediments to competition under a US/EU open skies regime could be the pervading influence of national interest, where member States of the EU could continue to ensure an "inside track" to their carriers, which may include insistence that nationals of a country fly on their national carrier; placement in a computer reservation system where interested parties could give prominence in the system to their carriers; and the use of excessive user fees to discourage foreign carriers.

As to whether there should be absolute, untrammeled competition within the Americas and between the Americas and Europe was a critical issue for the coming years. One suggestion was to crystallize a "convergence of regulatory principles" between Europe and the United States in competition by establishing a Transatlantic Common Aviation Area (TCAA). This concept, suggested by the Association of European Airlines (AEA) in a policy statement, put forward detailed and realistic proposals on how to bring about an ideal regulatory convergence between the European region and the United States, addressing three areas:

1. Matters in respect of which harmonization is necessary;
2. Those in respect of which convergence could take the form of mutual recognition; and
3. Those which could in principle be left at the discretion of each party.

The TCAA concept advocated the freedom of the parties to provide services; addresses issues pertaining to airline ownership and the right of establishment; provides recommendations with regard to competition policy; and offers guidelines on the leasing of aircraft.

Since the TCAA aimed at replacing traditional governmental regulatory control of such aspects of competition as market entry and pricing, the issues emerging from competition policy became by far the most complex and difficult to deal with, within the parameters of the TCAA. Although the fundamental postulates of competition in Europe (as followed through by European Union regulations) and the United States were broadly similar in intent, and both depend to a certain extent on the application of extra-territoriality in their regulations, there were obvious differences such as those embodied in the different approaches to transatlantic airline alliances. Also, the United States stringently relied on a principle of "public interest" in its air transportation policy, while European competition rules are not as explicit in their policies. The basic essence of a TCAA was therefore to establish the principle that matters of route sharing, capacity, pricing and frequency of services should be driven by market forces rather than be determined by governmental intervention. This way a certain commonality might be established between air transport of the two regions.

The United States had, over the past few decades, steadfastly advocated the need for open skies agreements with its partners in aviation. At the time, at the bilateral level, 38 "open skies" bilateral air services agreements had been concluded by 17 States or areas in the Asia Pacific region, with the United States being one of the partners in 12 cases. With Africa, the United States had concluded open skies agreements with 16 African States. While no U.S. carriers had been directly serving to Africa, they have expanded code share services with European carriers. Also, several African carriers inaugurated services to the United States.

For its part, The EU had also made progress in that the European Commission has concluded horizontal agreements so far with Australia, New Zealand and Singapore, all of which were initialled in 2005. The European Commission has also asked the Council of the EU to grant more comprehensive negotiating mandates for the creation of open aviation areas (OAAs) with Australia, China, and New Zealand.

With regard to Africa, the EU had achieved a significant level of progress. One of the European Commission's negotiating mandates conferred by the Council of the EU was to negotiate, on behalf of all member States, a Euro-Mediterranean aviation agreement with Morocco. This agreement was initialled in December 2005 and eventually replaced all the bilateral air services agreements between Morocco and the EU member States. The European Commission had also been conferred a horizontal mandate to replace certain specific provisions in the existing bilateral agreements declared contrary to Community law. In response to the European Commission's negotiating mandates, African Ministers agreed in May 2005 that it was necessary to adopt a common external policy and recommended to carry out a two phase plan of action for this purpose.

As can be noted, irrespective of the difficulties which arose from the transition from a traditional and entrenched bilateral method of negotiation, both the US and EU forged ahead towards their goal of open skies with an impressive list of precedent. The collective position of these two giants were rife with complex realities of competition and cannot be compared with other nations who might place open skies on a bilateral negotiation table and consider it a done deal if the other party accepts. Nor could the US/EU open their territories to unlimited and untrammeled open skies. There had to be a sense of direction where the two parties are headed when capacity, pricing and frequency are open. This direction should address the outcome of open skies and the various exigencies that might follow, such as complexities in slot allocation, national interest, possible carrier alliances, and secondary business stemming from open skies.

All inhibitors to open skies notwithstanding, the overall benefits of liberalization must outweigh the consequences of protectionism. As one commentator has said, when all is said and done, "every argument against open skies is an argument in favour of protecting some airline or other against competition…on the flip side of capacity dumping and predatory pricing you find a smashing deal for the markets in and out of the country, more business and tourist travelers, more goods moving by air, hotels flourishing, the overall economy better off and everybody's happy". If the US and the EU were to adopt this same philosophy they could have only ended the negotiations with a "win-win" deal. Which they did.

8.2 A Case Study of ASEAN and Indonesia

To say that Indonesia is an enigma in air transport is an understatement. On the one hand, the demand for air transport in Indonesia is higher in proportion to its GDP per capita. Its economy can be expected to grow 6–10 % annually. A single aviation market could add another 6–10 % growth in sheer demand. It is one of the wealthiest countries in the world, being the 16th richest country currently, and, according to an Airbus forecast, will be the 7th richest in 2030. Yet its airports are badly in need of expansion, its infrastructure is bursting at its seems, and above all, its airlines are strongly resisting liberalization of air transport in the region for fear of being wiped out by stronger contenders in the region. Against this backdrop, it is incontrovertible that Indonesia's civil aviation is intrinsically linked to regional and global considerations. Indonesia's archipelagic topography makes its people heavily reliant on safe, regular and reliable air services that may connect them not only internally but also to the outside world.

A single aviation market in the ASEAN region will bring both benefits to Indonesia and challengers to its air transport sector.

Air transport generated 1.8 million direct jobs and supported 24.2 million jobs in the Asia Pacific[1] region in 2012. The Industry generated $516 billion during the same period[2] and is therefore the region is a major player in global air transport.

The Association of South East Asian Nations (ASEAN) is comprised of ten countries[3] which will come under an Open Skies Policy (OSP) under the ASEAN

[1] Although the Asia Pacific region has been defined as a business region consisting of the whole of Asia as well as the countries of the Pacific Rim, there does not seem to be a clear definitive identification of the component States of the region. However, the 10 ASEAN nations are included in economic considerations of the region.

[2] *Aviation Benefits Beyond Borders*, Air transport Action Group (ATAG), April 2014, at 36.

[3] The Association of Southeast Asian Nations (ASEAN) was formed in 1967 by Indonesia, Malaysia, the Philippines, Singapore, and Thailand to promote political and economic cooperation and regional stability. Brunei, Cambodia, Laos, Myanmar and Vietnam later joined the Association. ASEAN covers a land area of 4.46 million km^2, which is 3 % of the total land area of Earth, and has a population of approximately 600 million people, which is 8.8 % of the world's population. The sea area of ASEAN is about three times larger than its land counterpart. In 2012, its combined nominal GDP had grown to more than US $2.3 trillion.

Single Aviation Market (ASAM) from 2015.[4] The ASEAN Economic Community (AEC),[5] which has as one of its aims the creation of a region fully integrated into the global economy, and which in turn is reflective of an economic integration plan in the region, considers air transport as a critical and significant thread in the transport fabric of the region. The implementation of ASAM would hinge heavily upon the disparity of the political and economic nature of ASEAN member States. One of the key factors in this equation is the differing priorities accorded by each State to air transport.

Indonesia, which is the largest Member State of ASEAN, is an archipelagic country embracing nearly 17,000 islands accommodating 231.4 million people which represents 39 % of the total population of the ASEAN member States. These islands span over 113,700 square miles. This spread of population inevitably brings to bear a compelling need to lay emphasis on and prioritize connectivity if the country were to prosper and grow its GDP[6] through the enhancement of resources and industries. Air transport is the logical driver of connectivity as other modes of transport such as road, rail and sea are cumbersome due both to the vast distances involved in Indonesia and the poor infrastructure that is incapable of supporting them.

Indonesia had 29 international airports in 2012, the largest in the ASEAN region.[7] The country has an open skies agreement with the United States (signed in 2004). In exchange for full fifth freedom rights into Indonesia for any US carrier from any point in the United States, Garuda Indonesia (as the only Indonesian carrier) operates to the United States. Garuda uses liberal code sharing agreements with other carriers to operate US-Indonesia and EU-Indonesia routes. 25 million international passengers—48 % of whom travel to and from ASEAN member

---

[4] Recognising the growing importance of trade in services, ASEAN countries officially launched their joint effort to work towards free flow of trade in services within the region through the signing of ASEAN Framework Agreement on Services (AFAS) on 15 December 1995 by *ASEAN Economic Ministers (AEM)* during the 5th ASEAN Summit in Bangkok, Thailand.

[5] The ASEAN Economic Community (AEC) is the goal of regional economic integration by 2015. AEC envisions the following key characteristics: (a) a single market and production base, (b) a highly competitive economic region, (c) a region of equitable economic development, and (d) a region fully integrated into the global economy. According to the AEC Blueprint which has been signed by all ASEAN nations, the AEC is the realisation of the end goal of economic integration as espoused in the Vision 2020, which is based on a convergence of interests of ASEAN Member Countries to deepen and broaden economic integration through existing and new initiatives with clear timelines. In establishing the AEC, ASEAN is required to act in accordance to the principles of an open, outward-looking, inclusive, and market-driven economy consistent with multilateral rules as well as adherence to rules-based systems for effective compliance and implementation of economic commitments.

[6] The Organization for Economic Cooperation and Development (OECD) defines Gross National Product (GNP) as an aggregate measure of production equal to the sum of the gross values added of all resident institutional units engaged in production (plus any taxes, and minus any subsidies, on products not included in the value of their outputs).

[7] The second largest number of international airports is 10 in the Philippines. Singapore and Brunei have one international airport each and no domestic airport.

States—reflect an average of 20 % traffic increase annually (both domestic and international traffic),[8] which is consistent with the rate of national economic growth.

ASAM is one of the key priorities in the overall AEC scheme and AEC will come into effect with the ASEAN Open Skies Agreement (AOSA) in 2015. ASAM is based on two supporting agreements—The Multilateral Agreement on Air Services (MAAS) which was signed in 2005 and the Multilateral Agreement on the Full Liberalization of Passenger Services (MAFLPAS) which was signed in 2010. Both these agreements provide for the relaxation of market access (which is provided for in the Protocols of both agreements) with regard to third, fourth and fifth freedom traffic rights, to the exclusion of seventh freedom rights[9] and cabotage.[10] There is also a great degree of relaxation of ownership and control of airlines, allowing operators flexibility to exercise their principle place of business as the main basis of their operations from and to ASEAN member States.[11] Furthermore, there is flexibility allowing a carrier to override the traditional 49 % requirement of foreign investment in an airline subject to the condition that the grantor State which receives the application to operate air services into its territory agrees.

However, ASAM is fettered by the fact that, for it to come into effect, at least three ASEAN member States have to deposit their instruments of ratification or acceptance, which dispels a common misconception that AOSA transforms the region into a fully liberalized air transport sector. Indonesia is expected to ratify the two MAFLPAS Protocols in 2015. The liberalization policy of Indonesia—particularly with regard to coming under AOSA—is resisted by the carriers of Indonesia who would prefer to stay with the *status quo ante* of the bilateral air services agreement that ensures them of some protection and control against the giants of the

[8] Five large airlines—Garuda, Lion air, AirAsia, Sriwijaya and Batavia operated 90 % of the international air routes in Indonesia in 2012. Batavia went bankrupt in 2013.

[9] A seventh freedom right grants the right to a country's aircraft to fly between two foreign countries while not offering flights to one's own country. It is the right or privilege granted by one country to another country, of transporting traffic between the territory of the granting country and any third country with no requirement to include on such operation any point in the territory of the recipient country. The service need not connect to or be an extension of any service to/from the home country of the carrier. For example, air traffic from England going to Canada on a U.S. airline flight that does not stop in the United States.

[10] Cabotage is the exclusive right of a country to operate the air traffic within its territory. This right is usually not given to foreign carriers. Called the Eighth Freedom, this right is not expressly forbidden but considered a discretionary right in the Chicago Convention which provides that each contracting State shall have the right to refuse permission to the aircraft of other contracting States to take on in its territory passengers, mail and cargo carried for remuneration or hire and destined for another point within its territory. Each contracting State undertakes not to enter into any arrangements which specifically grant any such privilege on an exclusive basis to any other State or an airline of any other State, and not to obtain any such exclusive privilege from any other State. Chicago Convention, *supra*, Chap. 1, note 5, Article 7.

[11] This concession admits of an airline's majority shares to be outside the State which designates the carrier, provided the seat of incorporation lies in the designating State.

ASEAN region, who they fear will not respect the market needs of the region.[12] The Indonesian carriers' argument is that they are sufficient to provide the capacity needed and that they are capable of catering to Indonesian traffic demands both internationally and domestically. Other member States, particularly Singapore and Malaysia are less rigid, and both Singapore and Malaysia have already an open skies agreement between them. One of the reasons why both Indonesian and Philippines' carriers oppose open skies is that they have substantial domestic markets which they hub in terms of connecting them to international transport which they think may be captured by the *carte blanche* entry into their markets by carriers such as Singapore Airlines, Thai Airways International and Malaysian Airlines System.

On the airports side, Indonesian airports could be affected by AOSA, as 16 of the 26 largest airports in the country are already running at over capacity. Sukarno-Hatta Airport in Jakarta has its existing terminals overflowing and its two runways are overcrowded. Although in 2013 a major upgrade of the airport was undertaken, it is expected that the exponential growth of capacity demand will make this upgrade ineffective by the time it is completed. Even if the overall upgrade of its airports are underway, Indonesia could have the common problem of capacity demand overriding any measures taken to meet it.

8.3 Integration into the Global Economy

8.3.1 The Global Economy

AEC's aim to integrate ASAM into the global economy (which has already been alluded to) is a key consideration in addressing the effects of AOSA on Indonesia. AEC aims at four key results: a single market and production base; a highly competitive economic region; a region of equitable economic development; and a region fully integrated into the world economy. These aims have already been recognized by the global aviation community. For example, The ASEAN Air Transport Integration Project (AATIP) of the European Aviation Safety Agency (EASA) aims at contributing towards sustainable ASEAN economic growth and the integration of the ASEAN Economic Community (AEC), through the development of the civil air transport sector.

In order to support this overall objective, AATIP will facilitate the development of the institutional frameworks to strengthen institutional capacities within ASEAN with a view to supporting the achievement of a safe, secure and sustainable ASEAN Single Aviation Market based on high regulatory standards.

---

[12] Most bilateral air services agreements between Indonesia and ASEAN member countries have stringent capacity and frequency limitations enforced by confidential memoranda of understanding. The most constrained routes are Jakarta-Singapore and Jakarta-Kuala Lumpur.

If AEC is to integrate into the world economy, the members of ASEAN would have to be aware of economic trends in global air transport and the challenges of market access worldwide. Therefore, a key consideration for Indonesia would be to align its aviation policy and approach towards global trends. The first step would be to forsake protectionism and embrace competition and the way forward towards connectivity. In this regard, a question for Indonesia would be whether ASAM would boost its tourism industry. UNWTO recorded in 2007 that the number of international tourist arrivals has risen from 25 million in 1950 to 840 million in 2006. The revenues generated by these arrivals—not including airline ticket sales and revenues from domestic tourism—have risen at an average rate of 11.2 % a year (adjusted for inflation) over the same span of time, which is nearly twice as fast as arrivals and a growth rate that far outstrips that of the world economy as a whole. International tourism receipts reached 735 billion dollars in 2006, almost 900 billion including air tickets, making tourism not only a socio-economic driver but one of the largest categories of international trade. Tourism represents one quarter of all exports of services—40 % with air transport revenues included. It is also noteworthy that the growth trend will continue, as according to a news release of 18 September 2007 issued by ICAO, total world airline scheduled passenger traffic in terms of passenger-kilometers is expected to grow at an average annual rate of 4.6 % up to the year 2025, half a percentage point lower than the growth rate achieved over the period 1985–2005, according to forecasts prepared by ICAO. Total freight traffic growth over the same period is forecast to be stronger, at 6.6 % per annum in terms of freight tonne-kilometers.

There can be particular benefits from tourism for the poorest economies, where international arrivals are growing at twice the rate in the industrialized States. Although they are still at a low level, the tourism receipts of the Least Developed Countries increased fivefold between 1990 and 2005, from 1 billion to 5 billion dollars. Tourism has become one of the largest sources of foreign exchange revenues for developing countries generally and for the 49 LDCs specifically, reducing their foreign debt and diversifying their economies. Tourism is often the principal service sector activity and it is a notably effective catalyst for gender equality, employment of young people, rural regeneration, cultural preservation and nature conservation.

Therefore WTO, which is the specialized agency of the United Nations with a central and decisive role in promoting the development of responsible, sustainable and universally accessible tourism, concludes that, for all these reasons, tourism can play a major role in improving the standard of living of disadvantaged populations and helping them lift themselves above the poverty threshold. Tourism can be a primary tool for achievement of the Millennium Development Goals of the United Nations, as long as a balance with climate change effects is maintained.

There is no doubt that tourism and air transport are symbiotic. Travel and tourism, the largest combination of industries and the largest creator of wealth, is estimated to generate $3.5 trillion a year in activity and potentially provides employment to 130 million people worldwide. This accounts for 10 % of the world's GDP, 10.3 % of the world's wages, 9.8 % of the profits and 11.7 % of indirect and direct taxes. WTO has recorded that about 40 % of the 840 million international tourist arrivals in 2006 were by air. In terms of long haul destinations, this figure could be even higher. Furthermore, the vast majority of the 931 million international passengers in 2006 were tourists.

In the context of ASAM, would ASEAN airlines consider alliances? Would Indonesian airlines benefit by entering into such discussions with their ASEAN colleagues? This question could be answered with a look at the global economy and the situation of the airlines. Tony Tyler, Director General of IATA, the association of world airlines, had this to say at the Annual General Meeting of IATA held in Beijing in June 2012, As reported in *Air Transport News*:

> the major benefits of aviation such as connectivity and economic welfares are evident in tremendous contributions of aviation to the global economy such as the provision of 57 million jobs worldwide. However, the state of the industry has been characterised as fragile, with 631 billion dollars revenue but only 3 billion dollar profit for 2012. with high oil prices, political risks around the world and the crisis in the Eurozone a large number of airlines is struggling to keep revenues ahead of costs. Governments and airlines should work as strong partners in order for modern economies to grow prosper and create jobs through being exposed to global opportunities.

Ironically, one could place these facts against the backdrop of pronouncements of high level policy makers who are increasingly reaching the conclusion that aviation is making a substantial contribution to the global economy. *Oxford Economics*, an economic forecasting agency, in a recent report recognizes the wide range of benefits that air transport brings to economies and societies globally.

This report, issued in June 2009,[13] also suggests that the world's future prosperity may depend on a growing and thriving aviation industry, which currently supports nearly 8 % of the world's economy, and questions the environmental benefits and social impacts of limiting that growth. One wonders, then, as to why an indispensable economic tool such as the air transport industry which contributes so substantially to the global economy,[14] is shackled by restrictions which other similar businesses are not subject to.

---

[13] Oxford Economics, *Aviation, The Real World Wide Web*, June 2009, http://www.oxfordeconomics.com/free/pdfs/ox_econ_aviation_report/main.html.

[14] ICAO has estimated that the direct contribution of civil aviation, in terms of the consolidated output of air carriers, other commercial operators and their affiliates, was 370 billion US dollars. Direct employment on site at airports and by air navigation services providers generated 1.9 million jobs while production by aerospace and other manufacturing industries employed another 1.8 million people. Overall, the aviation industry directly employed no less than 6 million persons in 1998. These direct economic activities have multiplier effects upon industries providing either aviation-specific and other inputs or consumer products. In simple terms, every U.S. $100 of output produced and every 100 jobs created by air transport trigger additional demand of U.S. $325 and in turn 610 jobs in other industries. The total economic contribution of air transport, consisting of the direct economic activities and the multiplier effects, is estimated at U.S. $1360 billion output and 27.7 million jobs worldwide. See The Economic Contribution of Civil Aviation, ICAO Circular 219: 2004 at 1. Also *The Economic Benefits of Air Transport*, 2000 Edition, ATAG:IATA, at 7. Also, *ICAO Circular 219, Id.*, at 4.

Strategic alliances of the airline industry is but a natural corollary to the exponential growth of international air transport as an industry. In an open skies environment under ASAM, alliances between disparately sized carriers could be a distinct possibility. The concept itself is based on the theory that with rapid demand for air transport, requiring a doubling of the 16,000 world aircraft fleet by the year 2015, these would be a compelling need for new connections between points and more frequencies to serve these connections. There is no stopping this trend, which has already swept the aviation industry. There is, however, one point of caution. The fundamental postulate of air transport has been, and remains to be, safety of passengers. The proliferation of aircraft in the skies may challenge airline safety, if parallel measures are not set in motion to ensure the safe passage of the thousands of aircraft in the sky.

How does one approach market access in the coming decades. At the most fundamental level, the advantages of free trade as would apply to air transport would be that it would encourage States to trade freely with their trading partners which would in turn help in the growth of the global economy; it would give the consumer a better choice of products and competition generated by free trade would bring down the price of the product. Arguments against free trade in air transport would be that globalization and liberalization will take jobs away from a State; the limit of imports would keep money in the State; free trade could be a threat to national security and a State could develop dependence on the expertise of other more advanced States. Free trade increases national wealth and promotes foreign investment, both of which are absent in the present structure of market access in many States.

The main consideration, leading up to efforts by the international aviation community to achieve a deregulated global airline industry, is involved with the question as to whether free market principles can be applied globally to air transport. What needs to be considered is whether we are ready to accept the throwbacks as consequences of free market competition in air transport, particularly in losing national prestige projected by flag carriers. One of the corollaries to industry deregulation is the introduction of free market competition when companies switch from operative performance to competitive performance. Competition therefore emphasizes the need to focus on a company's performance in relation to its competitors. This principle can be readily apply to various industries that have already been deregulated, such as the motor vehicle industry, chemical industry and information technology industry. The operative question is "are these good analogies for application to the air transport industry?" Whatever be the answer to this question, if the deregulated domestic air transport industry of the United States were to be considered an analogy, one could say that a deregulated system in the United States, introduced in 1978, has led to a more efficient airline system in the country. Whatever be the case, access to facilities in a competitive market is essential toward attaining fluidity of market forces. In the air transport industry, this can be translated to mean that if free markets do not exist in the supply of complementary facilities, there will be no positive impact of liberalization. The complementary services in the supply of air transport are airport access, computer reservation systems and airport and air regulation services.

The International Chamber of Commerce (ICC), in a policy statement has expressed the view that the efficiency of air transport would be enhanced by creating more open markets and more flexibility with regard to foreign ownership. Given air transport's capability to facilitate economic activity,[15] its liberalization would enable the sectors that make use of it to become non efficient. ICC was in favor of a freer exchange of air services throughout the world and is convinced that it is time to move beyond the existing bilateral system, toward a genuine multilateral liberalization of air transport. Of course, liberalization would give way to competition, which in turn would impel airlines to pool their resources in order that they maximize on such assets as code sharing and airport slots. However, alliances do not necessarily mean lack of competition between partners. Airlines within alliances have to do their utmost to gain market access and keep their businesses alive. In order to do this both private enterprises and the States in which these enterprises are entrenched have to be equally competitive.

Any agreement to bring in an aspect of trade within a liberalized framework is generally a pro-active measure, which brings to bear the willingness and ability of the governments to face trading issues squarely in the eye. However, any agreement for trading benefits would be ineffective without the element of competition, both between enterprises and between States. The essential requisite for success in trading relations is competition, which in turn leads to national prosperity. A free trade agreement is merely the catalyst in the process.

The strongest thrust of globalization in the business world is its ability to generate competition within and between nations to offer the best goods and services at the lowest prices. The quality of services and pricing in China as an off-shore base have encouraged other nations, such as Malaysia, Thailand, Ireland, Vietnam, Brazil and Mexico to vigorously compete as viable off-shore bases.

The World Is Flat: A Brief History of the Twenty-first Century[16] is a best-selling book which analyzes the progress of globalization with a focus on the early twenty-first century. It was first released in 2005 and later released as an updated and expanded edition last year in 2006. The author, Thomas L. Friedman uses the metaphor of a flat world suggesting that the competitive playing fields between industrial and emerging market countries are leveling. This is both an interesting

[15] Economics has confirmed that aviation's contribution to the global economy supports 57 million jobs and some $2.2 trillion in economic activity. Oxford Economics projected that aviation will grow about 5 % annually to 2030. That would see passenger numbers rise to 5.9 billion and cargo shipments could triple to nearly 150 million tonnes. This connectivity would support 82 million jobs and $6.9 trillion of global GDP. If growth is held back by even one %age point, the global economy would forfeit 14 million jobs and over $1 trillion in GDP contribution from aviation. Aviation's benefits are not guaranteed. Aviation is expected to grow about 5 % annually to 2030. If that growth is held back by even one percentage point, the global economy will forfeit over a trillion dollars and 14 million jobs.

[16] Farrar, Straus and Giroux: New York, 2006.

and real phenomenon. The woes of the air transport industry have always centered on the claim of most that aviation is effectively precluded from entering the free market due to government meddling. This seems to go contrary to the Keynesian view of the advantages of some degree of government control of the free market economy. Harsh restrictions on ownership and control of airlines and market access are stringent tools of government control. The argument for deregulation, as against government control, pervaded the economic crisis of 2008 where some argued that the bubble burst because the markets had not been free enough from government meddling and that the US Congress pushed lenders to have unfettered ability to lend to those who could not pay back their loans, leading to a housing bubble. Pankaj Ghemavat, referring to the 2008 bubble says:

> Media magnate Rupert Murdoch blamed the Government for the debacle, stating: "It's very easy to blame the free market but how did we get to the housing bubble? We got it because of Congress pushing Fannie Mae and Freddie Mac into lending money to people who couldn't afford it and blowing up the price of housing. . ..[17]

In the air transport sector this principle is reversed in that governments meddle in fettering the freedom of air carriers by imposing political and economic restrictions. This trend seems to continue, much to the disappointment of the air transport industry. Anne-Marie Slaughter of Princeton University speaks of a new world order where the State will not be the only actor in the international system but will still be the most important actor; the State will not disappear but will disaggregate into its component institutions, which will increasingly interact principally with their foreign counterparts across borders; these institutions will represent distinct national or State interests, even as they also recognize common professional identities; and government networks will exist alongside and sometimes within more traditional international Organizations.[18] For air transport this is good news in that the State will not be the only actor in the international system.

8.3.2 ASAM and Air Transport Economics in Indonesia

In the 1960s and 1970s, Indonesia followed a socialist ideology where air transport was exclusively conducted by two State owned enterprises: Garuda Indonesian and Merpati Nusantara airlines. The Ministry of Communications set regulations and there were no private air carriers that operated and there was no competition for the two carriers. The new order which followed under General Suharto was a mix between socialist and liberal ideology, introduced with Law No. 1 of 1967 which introduced foreign investments. This was followed by Ministerial Decree SK 13/6/ 1971 which allowed the creation of new airlines owned by private companies while the State owned Garuda International Airways controlled the trunk routes. An year

[17] Ghemawat (2011), p. 12.
[18] Slaughter (2004), p. 18.

earlier, Ministerial Decree 31/U/1970 allowed general aviation to operate under licence.

There was further relaxation under the regime of President Susilo Bambang Yudhoyono where private airlines grew. Regulation KM 81 of 2004 contained requirements for the establishment of new airlines followed by the *Law No. 1 of 2009 on Aviation* (The Aviation Law) which covered *inter alia* subjects such as sovereignty, aircraft operation, airworthiness of aircraft, safety, security and insurance. The Act also contains provisions on leasing, fares and liability of air carriers. However, it does not detail a clear policy on competition or air carrier access except to say that transparency and antimonopoly are the two guiding principles in aviation law in the country.

The philosophy of the 2009 Act is that it is not necessary to have too many airlines but existing airlines must be able to compete. This approach may place a burden on the main carriers operating at present but on the other hand, may also give them the impetus to compete with more versatile carriers in the region under ASAM. Airlines must have sufficient aircraft ownership; necessary capital investment; have a national majority holder; be supported by a bank guarantee and have the human resources capability to sustain the airline. New carriers must, in addition to providing a bank guarantee, prove that they own enough aircraft to support their operating licenses and hold a business permit. They must operate at least 10 aircraft (5 owned and 5 leased) if they are scheduled airlines and 3 aircraft (1 owned and 2 leased) in the instance of non-scheduled and all-cargo carriers.

Indonesia is in the Asia Pacific region which has grown the fastest among the regions in air transport. In particular, traffic within the region has grown very fast, according to an Airbus survey.[19] Asia Pacific's domestic traffic accounted for 36 % of global domestic traffic and it is envisioned that revenue passenger kilometres will reach 52 % by 2032. Aircraft fleets of carriers in the region are the youngest in the world. The Airbus study states:

> Projected future global network development in Asia-Pacific will have an important role in shaping the global network as a result of the dynamic air travel market in the region. Examining just the long haul routes, more than 50 % of the new routes created between 2013 and 2032 will be connected to Asia Pacific. This gives a flavour of the dynamic nature of air transportation in the region.[20]

In 2012, Indonesia was the 16th largest economy in the world, and it is forecast that in 2030 it will be 7th largest economy.[21] Then, it had 45 million members of the consuming class and in 2030 it will have 135 million. 53 % of the population were then in cities, which will increase to 71 % of the population in 2030 and the number of skilled workers will double, from 55 million to 113 million, increasing market opportunity from $0.5 trillion in 2012 to $1.8 trillion in 2030.[22]

---

[19] *Airbus – Global Market Forecast – Future Journeys*, 2013–2032, at 61.

[20] *Ibid.*

[21] Oberman et al. (2012), p. 1.

[22] *Ibid.*

Indonesia is up to date with its technical requirements for the operation of aircraft. The regulators are taking a very positive stance at enhancing its standards on the technical aspects of air transport. When one considers the doldrums Indonesia was in the year 2007, Indonesia has shown a strong come back with a conscious attempt to restore the industry's reputation. The Transportation Ministry has introduced a proactive system of safety rating, based on three levels: Level One indicating no serious issues; Level Two highlighting problems to be fixed; and Level Three enforcing the termination of flight operations. No airline at the introduction of this process came up to Level One ranking but at present all the country's airlines have reached Level One. The Ministry of Transportation insists on all airlines to be subject to the mandatory IATA Operational Safety Audit (IOSA) to be up to date with all operational management and control systems, particularly in view of new airlines coming into the business to meet capacity demand.

However, despite these encouraging facts and statistics, there are some challenges that Indonesian air transport may face with ASAM. One such challenge is with regard to aircraft type that will be used intra regionally by the larger carriers such as Singapore Airlines and Thai Airways International. In a liberalized market in South East Asia, aircraft type will be a considerable determinant of market share as the average passenger carriage in the region is about 150–200 passengers on a trip of around 400 nautical miles. This is much higher than in North America or Europe and therefore large aircraft, such as the A380 could be used effectively by the larger carriers at a good load factor on intra regional routes.

Another impediment for Indonesia in the global scene, and in particular the intra-regional ASEAN scene is that it is a closed economy with a tendency to be protective in the air transport field. As stated earlier, the apprehension in opening its skies is partly due to the tremendous domestic air traffic Indonesia generates, which offers potential to competitors in terms of connecting such traffic to international routes through sixth freedom carriage. To add to this disadvantage, Indonesian carriers do not offer expansive and regular services to Europe and North America compared to its ASEAN competitors, thus encouraging them to offer such connections on a sixth freedom carriage basis. One commentator refers to the Indonesian apprehension as follows:

> This is the familiar operating model of major "sixth freedom" carriers worldwide such as Singapore Airlines, Emirates, KLM, Korean Air, Turkish Airlines and Qatar Airways. In essence, "sixth freedom" hub operations depend simply on two factors – a geographically strategic "hub" airport in the centre of airline routes to serve as a transit stop, and unlimited third/fourth freedom rights to operate numerous "spokes" (hence, the alternative nomenclature of "hub-and-spokes" to describe the "sixth freedom").
>
> Due to the Indonesian carriers' relatively limited international operations, most travellers from Europe, North America and Northeast Asia travel into and out of Indonesia on foreign carriers' sixth freedom operations. The largest operator in this regard is Singapore Airlines, which channels these travellers through its hub, Singapore Changi Airport. In recent years, the highly successful Malaysian low-cost carrier AirAsia has also begun to transport the budget-minded segment of travellers in this same manner through its hub at Kuala Lumpur

> International Airport. The discomfort with such sixth freedom operations accounts for Indonesia's reluctance to accept the ASEAN agreements...[23]

Such dynamics have led the Indonesian government to propose only five points for an "open skies" policy—the major cities of Jakarta, Surabaya, Medan, Makassar and Bali.

Amidst the protests of the Indonesian carriers against open skies in the ASEAN region, and the cautious attitude of a "managed" protectionist stance to liberalization, the Indonesian authorities are sending a somewhat confusing and bullish message about the beneficial effects that ASAM might bring the country. One of the reasons for sanguinity could well be the growing Indonesian economy which will allow increasing numbers of the middle class to travel on business and tourism and the belief that the fast growing ASEAN region will be an added impetus in pushing demand for travel through ASAM.[24]

One commentator is of the view that Indonesia needs a general overhaul of its philosophy that would enable the country to embrace a global perspective and ensure its rightful place in the world scene:

> Traditionally, our nation's place in the world has been framed by its leading role in Southeast Asia, similar to how Brazil was seen in relation to South America, South Africa to Africa and India to South Asia. With the inclusion of these four regional leaders in the G-20, Indonesia must prepare itself to take on a role that goes beyond national and regional concerns to include leadership on complex global issues.
>
> Indonesia is expected to help shape the discourse on international issues — this makes the ability to bridge cultures critical. To this end, we need to move away from xenophobic sentiments that promote false perceptions of the world and perpetuate obsolete policies based on national insecurity. We need ideas that embrace globalization and place Indonesia in a position of influence. A solid group of foreign-educated citizens is essential to achieving this role.[25]

This may well be the case, but in Indonesia's defense, it must be stated that not all of Indonesia's air transport has been parochial. There has been a conscious effort on the part of airlines of Indonesia to expand their business activities and improve on their business models. In 2010, Garuda Indonesia revamped its branding strategy by redesigning its corporate identity and launching a new service concept 'The Garuda Experience'. This measure earned the airline the title 'Most Improved Airline 2011' and 'World's Best Regional Airline 2012' from *SkyTrax*, a UK-based consultancy for in-flight research services. In the low-cost carrier segment, Lion Air became the largest client of Boeing in the Asia-Pacific region.

This has prompted one commentator to say:

> It is just a matter of time until Indonesia finally has one of world's largest aviation industries. The country's economic strength and growing middle class will continue to

[23] Tan (2013), p. 12.

[24] Gusti Ngurah Irwan Dharmawan (2012), p. 21.

[25] Riady (2010).

make it a strong market for air travel, with improved safety and infrastructure investment only adding to the industry's potential.[26]

The above notwithstanding, the most troubling dichotomy faced by Indonesian aviation is that although Indonesia has a relatively high demand for air travel,[27] its airports are in dire need of expansion[28] and infrastructure development if they were to come up to the standards of airports in other ASEAN region airports. Law No 1 of 2009 allows private investment in airports and this concession must be encouraged and pursued. Airports both in both the western and eastern regions need expansion. Secondary cities must be designated as alternative hubs that could ease the 70 % of total traffic which now flows through Sukarno-Hatta Airport.

Although ASEAN's ultimate aim is to forge, through ASAM, a truly single aviation market and strengthen its position with regard to market access for its airlines in other regions of the world, this dream may not be realized by 2015. Indonesia, with its strong reservations on unlimited third and fourth freedom rights for the giants of the region under open skies, and its weakened competitive ability with the large and forceful marketers of the region such as Singapore Airlines and Thai Airways International, may yet show a reluctance, and even a refusal to accept ASAM without reservation. Another glaring disadvantage of ASAM is that, unlike in the European Union's single skies, ASAM does not have a regulatory infrastructure in place together with established laws for competition policy, state aid, consumer rights and other areas, making the ASEAN open skies look very much like a "free for all". This would give the Indonesians more reason to be apprehensive of open skies in the region.

Ironing out ASAM to ensure that all players have a level playing field and that air services be operated in a fair and equitable manner[29] will take time, but eventually, if a deal acceptable to everyone can be struck, the benefits of an ASAM may outweigh parochial and protectionist interests. To use the words of the United States Department of State:

Direct air connections bring substantial economic benefits. Open Skies agreements expand cooperative marketing arrangements, liberalize charter regulations, improve flexibility for

---

[26] O'Neill (2012). See http://webershandwick.asia/the-fall-and-rise-of-aviation-in-indonesia/.

[27] The World Bank has estimated that the population with disposable income forming the middle class in Indonesia will grow from 130 million people to 240 million by 2021. See Saraswati and Hanaoka (2013), p. 11.

[28] Sukarno-Hatta Airport served more than 50 million passengers in 2012, more than twice its design capacity.

[29] The Preamble to the Convention on International Civil Aviation provides, *inter alia* that Contracting States agree to certain principles that would ensure international civil aviation is developed in a safe and orderly manner and that international air transport services may be established on the basis of equality of opportunity and operated soundly and economically. See Chicago Convention, *supra*, Chap. 1, note 5.

> airline operations, and include provisions committing both governments to observe high standards of safety and security. These agreements also produce countless new cultural links worldwide.[30]

One can only be hopeful that beyond 2015, the Indonesian carriers may have built their bases to an extend to be able to compete with other carriers. In the end, ASAM is not about protecting national carriers but forging economic growth in the region. It is about running with the flow, opening the ASEAN region to the world and cooperating with the major aviation players such as China with whom the ASEAN nations (including Indonesia as a signatory) have signed an air transport agreement. It would be up to the Indonesian authorities to weigh the balance between the interests of its overall economy, the development of the country and the benefits that would accrue to Indonesians, against and the national pride that the Indonesian carriers would bring the nation by carrying more Indonesian across the world than do other carriers.

References

Ghemawat P (2011) World 3.0 – global prosperity and how to achieve it. Harvard Business Review Press, p 12

Gusti Ngurah Irwan Dharmawan I (2012) The effect of air transport on economic development in Indonesia. Paper submitted to the Erasmus School of Economics, Rotterdam, August, p 21

O'Neill M (2012) The fall and rise of aviation in Indonesia, October 9th, 2012

Oberman R, Dobbs R, Budiman A, Thompson F, Rosse M (2012) The archipelago economy: unleashing Indonesia's potential. McKinsey Global Institute, September, p 1

Riady J (2010) Top-flight foreign-educated graduates can help Indonesia go truly global, 10 January 2010, http://www.livinginindonesiaforum.org/showthread.php/6197-What-s-wrong-with-Indonesia-amp-How-to-fix-it

Saraswati B, Hanaoka S (2013) Aviation policy in Indonesia and its relation to ASEAN single aviation market. Proc East Asia Soc Transp Stud 9:11

Slaughter A-M (2004) A new world order. Princeton University Press, Princeton, p 18

Tan AK-J (2013) Toward a single aviation market in ASEAN: regulatory reform and industry challenges. Eria Discussion Paper Series, ERIA-DP-2013-22, p 12

[30] http://www.state.gov/r/pa/pl/159347.htm.

Chapter 9
The NAFTA Free Trade Model

9.1 Introduction

With such models as the Trans Pacific Partnership being negotiated between the United States and Pacific countries the concept of free trade agreements are of relevant to the subject of air transport. Although open skies agreements allow market access between signatory countries, air transport has not featured prominently in free trade agreements. An example is the North American Free Trade Agreement (NAFTA) which has not allowed air transport to be included in the context of the three partners United States, Canada and Mexico, it is relevant to consider the model from an academic point of view.

A new generation of international trade law commenced its existence on 1 January 1994 with the entering into force of the world's two largest free trade agreements: the North American Free Trade Agreement (NAFTA) between Canada, Mexico and the United States and the European Economic Agreement (EEA) between the European Community and the European Free Trade Area (EFTA) countries. Earlier, on 15 December 1993, the Uruguay round of the General Agreement on Tariff and Trade (GATT),[1] which was the genetic inspiration of NAFTA and EFTA, was successfully concluded, indicating the dawn of an era of

[1] GATT is a multilateral body established in Geneva on 1 January 1948 on coming into force of the General Agreement on Tariff and Trade (GATT) negotiated and signed by 23 countries. GATT functions as the principal international body concerned with negotiating reduction of trade barriers and with international trade relations. While being an organization to which member States belong, where they could use it as a forum in which they can discuss and overcome their problems and negotiate to enlarge world trading opportunities, GATT is also a code of rules which is calculated to liberalize world trade. The Uruguay Round is the 8th round of multilateral trade negotiations held by GATT so far, and by far, one of the most complex. This round of negotiations is assisted by the Group of Negotiators on Services (GNS) which the GATT established in 1986 to follow the services negotiations. The GNS has drafted a detailed agreement comprising 35 articles and five annexes. See Abeyratne (1994), p. 2, for a discussion of the history of multilateral trade negotiations held by GATT.

R. Abeyratne, *Competition and Investment in Air Transport*,
DOI 10.1007/978-3-319-24372-6_9

international harmony in issues of trade. Fifty years after the Bretton Woods Conference, the signing of the new agreement on 15 April 1994 relating to GATT—which is now called the World Trade Organization (WTO)[2]—at the ministerial conference at Marrakesh will not only seal the legal structure created by the Bretton Woods system which incorporated the International Monetary Fund, the International Bank for Reconstruction and Development (World Bank) and the GATT, but it will also usher in a new world trading philosophy for the twenty-first century.

NAFTA, which was signed by all three Parties on 17 December 1992 in Ottawa, Mexico City and Washington, creates a $6.5 trillion free trade area for 370 million people and incorporates principles that are calculated to achieve free trade between Canada, Mexico and the United States. In addition to pure free trade principles, NAFTA addresses environmental and labour concerns of the Parties and contains provisions which are aimed at considering these issues equitably.

NAFTA[3] derives its legal origin from GATT, Article XXIV which provides exceptional treatment for Customs Unions and Free Trade Areas.[4] The fundamental principle of GATT—which is embodied in Article XXIV—is its most-favoured-nation (MFN) treatment clause whereby each party to the agreement accords immediately and unconditionally to services and service providers of any other party, treatment no less favourable than that it accords to like services and service providers of any other country.[5]

In its treatment of free trade between the three State Parties, NAFTA has addressed six main issues, in addition to the theme of the most-favoured-nation treatment clause. They are national treatment, performance requirements, extraterritoriality, investment incentives, dispute settlement and related investment environments. With regard to investments, the main thrust of NAFTA lies in its provisions relate to national treatment. The basic principle of national treatment relies on a guarantee that an investor of one party to the agreement will be accorded the same treatment by another Party to the agreement as that which is given to nationals of the latter Party, as the possibility of discrimination against parties

[2] Although the agreement is titled "Agreement Establishing the Multilateral Trade Organization" and establishes the Multilateral Trade Organization (MTO) in Article 1, with references thereafter in the agreement to the MTO, It is considered that the new name for GATT is "World Trade Organization". See *International Legal Materials*, Vol. XXXIII, No. 1 January 1994 at 2.

[3] The North American Free Trade Agreement Between The Government of the United States, Government of Canada and the Government of the United Mexican States (1992), *International Legal Materials*, Volume XXXII, No. 3 May 1993 at 612.

[4] See the *Text of the General Agreement on Tariffs and Trade*, Geneva, July 1986 at 41. The only major difference between these two instrumentalities of trade is that whereas in a Customs Union member States have to erect a common tariff wall towards the outside world, in a Free Trade Area member States are free to maintain or modify independently their external structure of tariffs and other barriers to imports from third countries.

[5] See *International Legal Materials Vol. XXXIII No. 1 January 1994 supra* note 3 at 34–38 (this chapter) for an interpretation of the most-favoured-nation treatment clause in the GATT Agreement of 1994 in relation to Free Trade Areas.

would violate the very principle of a liberalized investment regime. This does not mean however, that acceptable deviations from this principle are not necessarily encompassed in an agreement. Such exceptions are couched in precise terms in order to limit interpretations that may be contrary to the spirit of the accord in question. The second issue relating to performance requirements[6] has made NAFTA delimit the extent to which such requirements are permissible and the purposes for which these requirements exist. The third issue of extraterritoriality deals with laws and policies governing the transnational corporation and its internal regulations and directives and the extent to which these two categories of policy would conflict with the laws, regulations and practices of the host country, since there are no general principles of international law governing the application of laws and policies extraterritorially.[7] The fourth issue, which is investment incentives, addresses the sensitive and broader issue of subsidies. Investment incentives involved for the negotiators of NAFTA the prospect of including provision for duty draw backs for re-exported goods within the free trade area. The fifth issue—dispute settlement—has been resolved by Article 2005 of NAFTA which accepts the GATT dispute settlement principles as applicable to disputes arising out of the implementation of NAFTA. According to this provision, disputes may be settled either in GATT (when such disputes arise under a GATT principle) or in a successor to GATT as the case may be. The sixth issue, which is investment environment, deals with the plethora of laws applicable generally to antitrust laws, taxation, intellectual property laws, policy on competition and the control of investments, which may, in certain circumstances, be applied in a discriminatory fashion.

NAFTA has seemingly succeeded in bringing the above issues cohesively together in a single document to the mutual satisfaction of the three Parties concerned. This article will examine the economic reasons which convinced Canada, Mexico and the United States to enter into NAFTA. It will also identify and analyze the legal principles governing the agreement.

9.2 Economic Environment of NAFTA Parties

9.2.1 Canada

Canada has prolific foreign ownership in the private sector of the country's economy and therefore has approached the regulation of foreign investment somewhat

[6] For a detailed analysis of performance requirements see, Graham and Krugman (1989), p. 127.
[7] See Dallemeyer (1989), p. 565. See also generally Castel (1989).

cautiously.[8] This situation and the trends that are the corollaries of it have kindled the regulation of foreign investment, particularly with the help of appropriate measures that have been adopted.[9] With the establishment of the Foreign Investment Review Agency (hereafter FIRA),[10] Canada adopted a restrictive policy on foreign direct investment, screening every foreign investment through FIRA, regardless of resources offered and wealth and size of the investment, and allowing only foreign investments of significant benefit for Canada.[11]

FIRA was subject to sustained criticism from all parties who had dealings with the Agency. Complaints concerned the inordinate delay in the completion of process (which usually took from 12 to 18 months to process), its indiscreet screening process, the lack of care in considering relevant information, the seemingly arbitrary selection criteria and above all the fact that foreign investors were required to negotiate with an Agency, which in the end, was not the one having powers to take the final decision with regard to the investment proposal under consideration. FIRA was therefore perceived by Canada's trading partners as a discretionary organ impeding the free flow of trade and investments.[12] One of the examples of discontent among Canada's trading partners is seen in the formal complaint brought by the United States against certain requirements imposed by FIRA on foreign investors to the GATT in 1982, resulting in a GATT panel concluding that domestic sourcing requirements of the Agency were inconsistent with Canada's obligations as stipulated in Article III:4 of the GATT.[13] However, FIRA does not appear to have been as ominous in its negative approach to foreign

---

[8] See the *Gordon Commission Report* which analyzed the extent of foreign ownership in Canada. This Report proposed that foreign investment in Canada be controlled in four ways, including control by a central review board. *Report of the Royal Commission on Canada's Economic Prospects*, Queen's Printer:Ottawa, 1958. A subsequent study, *the Gray Report*, concluded that it was necessary to regulate foreign investment in Canada in view of the very high level of foreign ownership in Canada. According to this report 76 % of the energy sector and 90 % in various other sectors were under foreign ownership. See *Foreign Direct Investment in Canada*, Information Canada:Ottawa, 1972.

[9] For material on Investment between Canada and the United States see, MacDonald (1972), p. 71; Frank and Gudgeon (1975), p. 76; McDowall (1984); Wex (1984); Dewhirst and Rudiak (1986), p. 149. For material on investment similarities in Canada and Mexico see Hodgson (1975), p. 56.

[10] *Foreign Investment Review Act*, S.C. 1973–1974. See also Turner (1983), p. 333; Dewhirst (1984), p. 27.

[11] *Foreign Investment Review Act*, *supra* note 10 (this chapter), Article 2(1). Small investments of $5 million which employed no more than 200 employees however, required only the formality of registration.

[12] See Albrecht (1984), p. 149.

[13] Article III.4 of GATT provides:

The products of the territory of any contracting party imported into the territory of any other contracting party shall be accorded treatment no less than that accorded to like products of national origin in respect of all laws, regulations and requirements affecting their internal sale, purchase, transportation, distribution or use. The provisions of this paragraph shall not prevent the application of differential internal transportation charges which are based exclusively on the economic operation of the means of transport and not on the nationality of the product.

investment as it was perceived to have been, since the average approval rate of investment proposals was at that time 90 %. It is worthy of note however, that the level of foreign ownership in Canada declined from 37 % in 1971 to 24 % in 1987 under FIRA.

In 1984 Canada implemented a major reversal of the existing *status quo* by adopting its "Open for Business" Policy. The newly elected Conservative Government and its enactment of the *Investment Canada Act*[14] (hereafter ICA) ensured the repeal of the FIRA, replacing it with the Investment Canada Agency and an altogether new philosophy.[15] The ICA has conceptually encouraged foreign and domestic investment since basing its philosophy on the legislative recognition that foreign investment was beneficial to Canada.[16] The Act abolished the mandatory review process for the establishment of all new Canadian businesses and required a review in the case of direct acquisitions of $5 million or more, indirect acquisitions of over $150 million, and those investments relating to the advancement of culture.[17]

As a response to FIRA, ICA's investment review process was streamlined and given a lower profile than that enjoyed by FIRA. Under the new system, the investor received an answer to his investment proposal within 45 days of his proposal being deposited.[18] ICA adopts the criterion of selection to be that the reviewed instrument must be of net benefit to Canada.[19] This criterion has not impeded the successful review in any of the reviewed transactions brought before Investment Canada since its creation.[20]

See also, *Administration of the Foreign Investment Review Act*, Report of the Panel, July 25, 1983, BISD 30th Supp., at 140 (1982–1983).

[14] *Investment Canada Act*, R.S.C. 1985, c 28 (1st Supp.).

[15] See generally Heinz Juergen (1986). The author analyses comparatively the FIRA and ICA in his thesis. See also, Baker (1984), p. 47. Also generally, Spence (1986), p. 507.

[16] *ICA, supra* note 14 (this chapter), Article 2.

[17] *Ibid.* Article 14(1)(a). This regulation obviated the review of 90 % of the transactions that were required to be reviewed by FIRA. See however, Glover et al. (1985), p. 98 where the authors claim that the 10 % of the investment categories that were retained from the FIRA days comprised 90 % of the total transactional value of investments in Canada and therefore this apparent achievement of the ICA is merely illusory.

[18] *Id.*, Article 21(1). This period is extendable to a maximum of 75 days, unless the investor agrees to the contrary. Both the FIRA and the ICA considered the review of a proposal to be based on policy rather than law. See Arnett (1985), p. 26.

[19] *Id.* Article 21. The criteria determining "net benefit to Canada" are similar to those followed by FIRA and vests the Minister responsible for the implementation of the ICA with a large discretion. Article 20 provides that these criteria should be based upon the effect of investment on labour and employment, productivity, human and other resources, technological development and benefit, national and international market competition, the extent of participation by Canada and the consistency of the investment with national, economic and cultural policies of Canada.

[20] Three quarters of the 5266 investment proposals received up to the end of 1990 were not subject to review, and the 25 % of those that were reviewed had all been approved. See Investment Canada, *The Opportunities and Challenges of North American Free Trade: A Canadian*

There were also restrictions on foreign ownership in Canada in specific sectors of the economy, as is the case in many other countries. These sectors included aviation,[21] Banking,[22] Fisheries,[23] Agriculture,[24] Shipping,[25] oil, gas and uranium production[26] and broadcasting and cable distribution.[27]

9.2.2 The United States

Unlike in Canada, foreign investment has caused much less political upheaval in the United States. The United States has been, up to the 1980s, a strong net exporter of capital and did not have much to fear of foreign influence in trade issues within the country. In the earlier 1970s however, with the burgeoning oil crisis, petro dollars were invested in the United States leading to a 38.3 % and 22.3 % increase in foreign direct investment in 1973 and 1974 respectively. This figure can be contrasted with a 6 % increase on average in the decade between 1962 and 1972. This quantum leap in foreign investment caused grave concern in Congress and led to a national inquiry being instituted which called for measures affecting foreign investors in the United States. The 27 volume and 9000 page long final report, describing comprehensively the regulations facing foreign investors, concluded that a sufficient number of sectors were appropriately regulated, recommending that substantial change to the existing policy was unnecessary.[28]

Unlike Canada's comparably restrictive foreign investment policy, The United States has a liberal "open door policy" on foreign direct investment and is one of the most open economies in this respect. The United States International Investment Policy Statement of 1983 confirmed:

Perspective, Ottawa: Investment Canada, May 1991 at 44. See also Investment Canada, *Annual Report 1990–1991*, Investment Canada: Ottawa, 1991.

[21] *Aeronautics Act* 1985 R.S.C. Ch. F-2.

[22] *Bank Act* 1985, R.S.C., Ch B-1.

[23] *Fisheries Act* 1985 R.S.C., Ch F-14.

[24] *Western Stabilization Act* 1985, R.S.C. Ch W-7.

[25] *Canada Shipping Act* 1985 R.S.C. Ch S-9.

[26] *Territorial Lands Act*, 1985 R.S.C., Ch F-7; *Canadian Petroleum Resources Act*, 1986 Can. Stat., Ch 45; *Canada Oil and Gas Act*, S. C. 1981 C 81. The Canadian energy policy has often been subject to criticism on the basis that the Canadian Government adopted a policy of inordinate appropriation and that such policy was calculated to canadianize the industry. See Olmstead et al. (1984), p. 439. Also, for a descriptive analysis of Canada's Energy Programme see Mendes (1981), p. 475.

[27] *Broadcasting Act* 1985 R.S.C. Ch B-1.

[28] *Foreign Direct Investment Study Act of 1974*, 15 U.S.C. 786 (1982); *Foreign Direct Investment in the United States: Report of the Secretary of Commerce to the Congress in Compliance with the Foreign Investment Study Act of 1974* (1976).

> The United States has consistently welcomed foreign direct investment in this country. Such investment provides substantial benefit to the United States...We provide foreign investors fair, equitable and non-discriminatory treatment under our laws and regulations. We maintain exceptions to such treatment only as far as necessary to protect our security and related interests which are consistent with our international legal obligations.[29]

The United States has adopted the approach that the absence of regulation encourages investment and is beneficial to the United States economy,[30] and has therefore generally adopted a non-discriminatory treatment for foreign investors.[31] A commentator adds:

> Foreign nationals and companies are treated as favourably as nationals or companies of the United States with respect to the establishment and operation of enterprises in this country... Further, on the basis of the national treatment principal investors from other countries can generally make investments in this country on the same legal terms as American investors.[32]

The "open door policy" and national treatment principal however, does not reflect an accurate picture of the status of foreign investment in the United States. In contrast to what the "open door policy" is perceived to be, there are numerous laws that effectively preclude this policy from taking full effect, thus impeding foreign investments in the country. An example of this inhibitive approach is the 1988 "Exon-Florio Amendment" which provided the President with broad powers to review investments of foreign investors on his own initiative for any reason including those which directly or indirectly affect national security. The President may also review a foreign investment following the complaint of a third party.[33]

Investors may, by their own volition, serve notice on the Committee on Foreign Investment in the United States (CFIUS). In addition, the CFIUS can also decide to inquire into an investment by itself. It then advises the President of its decision with regard to an investment and the President ultimately decides whether or not the investment is contrary to national security interests. The notion of "national

[29] International Investment Policy Statement 19 *Weekly Comp. Pres. Doc. 1214* (September 1983) cited in Raby (1990), p. 400.

[30] See generally, Graham and Krugman (1989).

[31] The United States is limited by its Constitution and by the treaty provisions governing the country in its capacity to regulate foreign investment. There are built in guarantees that are offered to foreign investors in the *Friendship, Commerce and Navigation Treaties* (FCN), the *OECD Code of Liberalization of Capital Movements*, which have a direct effect on the United States legal system, and the guarantee of due process and non discrimination entrenched in the United States Constitution.

[32] Bale Jr. (1985), p. 207.

[33] *Omnibus Trade and Competitiveness Act of 1988*, Pub. L. No. 100-418, 102 Stat 1107 (West Supp. 1989) See Knee (1989), p. 475. Also, Sandstrom and Coccuza (1989), p. 65; Prichard (1988), p. 95.

security" is ambivalent in the context, lacking a precise definition, and this hiatus has left great discretion to the Executive in the implementation of the Act. The Act is used infrequently and each case is evaluated individually.[34]

In addition, many other sectors which operate through a fixed maximum level of foreign participation are excluded from foreign investor participation entirely or partially. These restrictions are seen at the federal level in the fields of communications,[35] transportation,[36] aviation,[37] energy and national resources,[38] banking[39] and defence.[40] Federal laws such as antitrust regulations contained in the *Clayton Act* of 1981[41]—which prohibits direct or indirect acquisition of shares of a company when it would affect or lessen competition in such a way as to incline the creation of a monopoly—and the *Sherman Act* of 1981[42]—which prohibits monopolies created by contract, conspiracy or other ways that restrain trade—may also bring to bear serious effects on the foreign investors' investments in the United States.

9.2.3 Mexico

Any trading enterprise conducting business in Mexico is subject *in limine*, to all laws and regulations applicable in Mexico, including environmental[43] and labour

[34] On 1 February 1990, President Bush issued an order based on the Exon-Florio Amendment, to the China National Technology Import and Export Corporation to divest its holdings in MAMCO, a United States manufacturer of aircraft components. MAMCO was a firm which fabricated custom made metal components for the use of manufacturing civilian aircraft and helicopters. See Mendenhall (1991), p. 294. See also Nance and Wasserman (1990), p. 926 for a discussion of Presidential powers on the regulation of foreign investments on national security grounds.

[35] The Federal Communications Commission may refuse to grant a broadcasting licence if the corporation applying for the licence is owned by foreigners and if it is in the national interest to refuse the grant of such licence. See *Federal Communications Act*, 47 U.S.C. 734 (1976).

[36] *Jones Act of 1920*, 46 U.S.C. 802 (1976 & Supp. V 1981).

[37] *Federal Aviation Act*, 49 U.S.C. 1301 (1976 & Supp. V 1981).

[38] *Atomic Energy Act*, 42 U.S.C. 2133, 2134 (1982), *Mineral Leasing Act*, 30 U.S.C. 22 (1982), *Agricultural Foreign Investment Disclosure Act*, 7 U.S.C. 3501 (1978). See also Scarborough (1985), p. 94.

[39] *National Bank Act*, 12 U.S.C. 1974. The restrictions aimed at under the banking laws of the United States are based on the nationality of the directors of a foreign investment company and not on foreign ownership.

[40] One of the most compelling elements of State control of foreign investment in the United States is based on national defence under the *Defence Production Act of 1950*, 50 U.S.C.A. App. 2170 (West Supp. 1989).

[41] *Clayton Act*, 15 U.S.C. 1981.

[42] *Sherman Act*, 15 U.S.C. 1981. See Davidow (1976), p. 895; Swan (1984), p. 177; Kraft (1988), p. 515.

[43] *Law to Achieve Ecological Balance and to Protect the Environment by Preventing and Controlling Pollution of the Atmosphere*, D.O., November 1988; *Law to Protect the Ecology and

laws.[44] The regulatory philosophy applicable to business conducted in Mexico follows a policy of internal growth in order to achieve self reliance and freedom from dependence on industrialized countries. In order to follow this philosophy, Mexico has, in the past, regulated heavily on foreign enterprises conducting business in Mexico, with the exception of foreign exchange controls. One of the forerunners of this regulatory structure is the *Foreign Investment Law*, which admits of tighter control of businesses, thereby ensuring maximum benefits to Mexico. Investors have to go through a stringent review process and are curbed to a maximum of 49 % of participation. They were also required to operate their businesses strictly within the commercial restraints imposed by Mexico. This policy worked in the period 1950–1970, when food exports acted as a catalyst to Mexico's commercial policy. The period which followed Mexico's burdensome public debt owing to this policy called for the creation of public enterprises, to buttress the crisis laden agricultural sector and supplement inadequate oil revenues which could not *per se* sustain Mexico's economic program.[45] The Government of President Salinas therefore obviated the stringent regulatory trade policy in order to align Mexico to industrialized policy trends. Foreign investors are now given the opportunity to take advantage of Mexico's *Programa de Intercambio de Deuda Publica por Capital* or *Debt-for-Equity-Swap Program*[46] which has been designed to allow a foreign investor to buy a part of the debt owed by Mexico to private banks at a discounted rate of 40 %. The foreign investor may then liquidate the value of the debt in pesos through the Mexican Government so that he could finance his local investments. A similar program had been in existence since 1986 but was suspended in early 1988 as the increased cash flow caused by the influx of foreign investors caused further inflation in the country.[47]

The system of investment which followed operated on an auction basis, where a foreign investor secures a deposit and presents a detailed bid, proposing the nature of investment and the value of debt. This program is also available to public enterprises being privatised. The bidder who secures the contract is announced

---

Environment from Dangerous Substances, D.O. November 1988. These laws are implemented and enforced by the Secretariat of Urban Development and Ecology (SEDUE).

[44] See Free Trade Negotiations with Mexico: Environmental matters (1991) 3 *Int'l Environment Affairs*, 219. See also Mumme et al. (1988), p. 7; du Mars and Beltran el Rio (1988), p. 787 for an overview of environmental legislation in Mexico. See also Labour Canada, *Comparison of Labour Legislation of General Application in Canada*, the United States and Mexico, Labour Canada: Ottawa, March 1991 for an analysis of the labour laws of Mexico. For NAFTA issues on the labour and environmental position of Mexico see McCambly (1991), p. 47; Reyes (1991), p. 241.

[45] See generally, Kate (1978).

[46] *Programa de Intercambio de Deuda Publica por Capital*, D.O. March 30, 1990. This programme is an integral part of the understanding between Banks in Mexico and the Mexican Government which resulted from the *Brady Plan*. Banks have undertaken to write off Mexico's medium and long term debt, while providing new loans to the government and reducing interest rates.

[47] See Baker (1988), p. 333.

within 60 days of the closing of the bidding.[48] The U.S. $3.5 billion which were set aside for the auction program was sold out in two auctions. This was more encouraging than envisaged and the government has not decided to apply this programme to other debts.

The Mexican Constitution prohibits foreign ownership of lands in a zone which represents close to half of the country's entire surface. This zone stretches for 100 km along the borders and 50 km along the shores of the country. Owing to this prohibition, foreign investment in the tourism and hotel industry has been circumscribed. This constitutional impediment has been subject to much interpretation, one of which admits of foreigners holding beneficiary rights through a Mexican trust, in which a Mexican bank acts as trustee. This mechanism has been used for a sustained period of time, leading to its legal recognition since 1971. The first legal allowance made was in the nature of a presidential decree which allowed a fiduciary ownership by foreigners for a maximum period of 30 years.

The *Foreign Investment Law* which came into effect in 1973 provided for conditions and requirements for the establishment of a trust and required the Foreign Investment Commission to institute guidelines governing the criteria and procedures by which such trusts could be established.[49] The Ministry of Foreign Affairs had final power to approve a project on a case by case basis. The Mexican financial institution acting for a foreign investor, must apply for the permit on behalf of that foreign investor who wishes to buy tourism or industrial facilities within the prohibited zone. Once permission is granted, a certificate of immovable participation is issued to the foreign investor. This certificate does not grant the beneficiary rights in any real property on the land or in the building he uses. The certificate, which is valid for 30 years, cannot be amortized.[50] The foreign investor was however, given the option of leasing the premises from the trustee for a maximum period of 10 years. At the end of the period, all property rights of *utendi* and *fruendi* (use and enjoyment) on the premises were transferred to a Mexican national.[51]

The 1989 regulations introduced by the Salinas Government did not change the Constitution. They merely gave beneficiaries of land trusts the right to renew the certificate for periods of 30 years by establishing consecutive trusts.[52] Renewal was granted if the *Foreign Investment Law* and the first obligation of the foreign corporation were respected. The new regulations also provided for an issuance of permits from the Ministry of Foreign Affairs for investments admissible without

[48] Torres Landa (1990), p. 734.

[49] *Ley para Promover la Inversion Mexicano y Regular la Inversion Estranjera* D.O. March 9, 1973. The text of this law was reproduced in (1973) 12 *International Legal Materials*, 643.

[50] According to Article 20 of the *Foreign Investment Law, supra* note 49 (this chapter), there was no option to renew this permit.

[51] *Ibid.* Article 20.

[52] *Ibid.*

review from the Commission and for *Maquiladoras*.[53] The land trusts system does not seem to have discouraged foreign investment in Mexico since, a 30 year renewable trust is in most cases more advantageous than being a tenant.

During the era of strict restriction of foreign investment, the import of any product into Mexico required a licence. This requirement ensured the implementation of Mexico's economic policy of internal growth and import substitution and made access to the Mexican market for imported products difficult. Importation involved high customs duties and therefore, the level of imports was very low.[54] When Mexico acceded to GATT in 1986, the decision was taken to adhere to most of GATT's principles and incorporate them in Mexican law.[55] By a law of 1988, the import licence scheme was abolished.[56] Consequent to this measure being taken, only agricultural and food products such as wheat, powdered milk, cheese, coffee, sea food, and automobile industry products—approximately 3 % of imported goods into Mexico—required an import licence.

The foreign investor is not bound by the *Foreign Investment Law* to limit himself to domestic sources in the purchase or use of raw materials. Mexico has made concrete efforts to harmonize import classification by becoming party to the *United Nations Convention on Contracts for the International Sale of Goods*[57] and giving the foreign investor in Mexico an environment favourable to obtaining external goods necessary for the conduct of his business.

The only other element the foreign investor requires to comply with in the operation of a business in Mexico is the law relating to foreign exchange. Mexico, which has a history of liberal exchange control laws which required no exchange control from 1929 to 1982, had to enact a decree on exchange control in 1982.[58] This decree sought to eliminate the outflow of hard currency and to obtain more binding effect of small investment arrangements. This law however, was effective only for a few months before being superseded by the *Decreto Control de Cambio*.[59]

---

[53] *Maquiladoras* are in-bond assembly plants where only a fraction of their production is allowed to be sold on the domestic market. The parts are imported into Mexico, to be assembled and then exported to other countries. See Article 17 of the *Foreign Investment Law*, *supra* note 49 (this chapter).

[54] It is noteworthy that in the first 5 years of the decade between 1980 and 1990, imports into Mexico amounted to an average value of U.S. $12 billion per year whereas in 1990, imports amounted to a value of U.S. $32.8 billion.

[55] Mexico adhered to GATT's *Antidumping Code* of 1979, *Code on Import Licences* of 1979, *Code on Customs Valuation* of 1979 and the *Regulations Against Unfair Trade Practices*.

[56] *Ley del Impuesto General de Importation*, D.O. February 12, 1988.

[57] *U.N. Doc. A/CONF.97/18, Annex 1*, reprinted in (1980) 19 *International Legal Materials* 668.

[58] *Decreto que establece el control generalizado de cambios*, D.O. September 1, 1982. The imposition of foreign exchange control was exceptional in the context that the peso had not been subject to control.

[59] *Decreto de Cambio*, D.O. December 10, 1982.

This decree diversified the existing *status quo* of generalised control of trade to a scheme of restricted control, creating two foreign exchange markets: a controlled and a free market. The former, which was operated by banks, was regulated by government and had restricted access. The latter was an untrammelled free market based on the principles of supply and demand. The goal of this dual system was to replenish Mexico's treasury with as much foreign exchange as possible and use this resource where it was most needed. The free market alternative, also known as "coyote", has been designed to eliminate the black market and give people an alternative way to deal with their needs pertaining to currency.

The controlled market is regulated so that specific types of foreign currency income must be exchanged on the controlled market such as income from the export of goods,[60] and the funds used by maquiladoras to cover local costs and loans received from banks. This inflow of foreign currency which is directed to foreign banks, is used for the repayment of debts in foreign currency; the purchase of merchandise imports and payment of royalties and technical assistance and compliance with other contractual obligations. While the other controlled market uses foreign currency influx for debt servicing, the free market is used for pure currency operations. Since the difference between the exchange rates is approximately 2 %, the foreign investor does not necessarily find himself restricted by the above regulations. He can use the free market for all capital operations, investments, remittance of dividends, payment of royalties and intercompany loans. Furthermore, there are no exchange control regulations for travellers.

9.3 Historical and Economic Aspects of NAFTA: Background and Rationale

The *Free Trade Agreement* (FTA) between Canada and the United States which became effective on 1 January 1989 was the precursor of NAFTA and provided for the gradual elimination of tariffs and non-tariff barriers between the two countries. This elimination of tariff and non tariff barriers applied to goods and services, liberalization of investment provisions, the lessening of cross border business travel formalities and introduced an impartial dispute resolution process related to trade issues. The implementation of these objectives was to be achieved over a period of 10 years. This agreement was also broad and all encompassing, addressing most trade issues between Canada and the United States, while providing allowance for negotiations on unresolved issues such as questions on subsidies.

[60] This does not however, apply to exports of small value nor to the *maquiladora* industries. It is also noteworthy that the Mexican exporter is legally obliged to require that exports be paid for in a foreign currency.

The achievement of free trade between the two countries was a result of well though out negotiations and sustained interest.[61] Canada's main interest was to secure access to the United States market which supported two million Canadian jobs; prohibit discrimination in trade affairs; create a cohesive and binding tribunal of inquiry into trade issues and harmonize and facilitate commercial regulations involving business migration. The United States was interested in securing access to the Canadian market and persuading Canada to liberalize Canadian investment policies.

It is incontrovertible that the FTA played an effective role in bringing Canada and the United States towards the creation of NAFTA. Although the FTA was designed as a bilateral agreement, a provision of the Agreement provides for either party a right to enter into other free trade agreements. The application of the FTA to other parties was however, not included in the Agreement[62] on the basis of the *res inter alios acta nec nocere nec prosunt* principle which excludes the Agreement's binding effect on non Parties to the Agreement. The FTA's aim was limited specifically to trade relations between Canada and the United States.

Initially, an FTA between Mexico and Canada and/or the United States was not even feasible. However, the successful negotiation of the FTA between Canada and the United States seemingly provided inspiration for Mexico to pursue the philosophy that now obtains in the country—that of liberalised trading policy. At first, when the idea of NAFTA surfaced, Mexico's thinking was to enter into a bilateral treaty. This view was shared by the United States which sought the official endorsement: of a comprehensive bilateral Free Trade Agreement as the best vehicle to strengthen bilateral economic relations and meet the challenges of international competition.[63]

This proposed agreement was primarily designed to obviate tariff and non-tariff barriers, ensure protection of intellectual property rights, expand investment, trade and services and establish a dispute resolution mechanism. It was also agreed as a parallel objective to include discussions on labour and environmental issues.

The United States Congress allowed the NAFTA principles which had now evolved through FTA principles to be negotiated through a fast track procedure. Congress approval was conditional upon the institution of parallel talks on environmental and labour issues. One of the significant developments in the field of environment protection relating to NAFTA in the United States was the District Court ruling by Justice Ritchey that the United States administration was in breach of the *National Environmental Policy Act* by not drafting an environmental-impact

[61] On four previous occasions, *i.e.* in 1854, 1911, 1925 and 1948, Canada and the United States had discussed introducing free trade between the two countries. In the case of Mexico however, NAFTA was the first instance that Mexico considered introducing free trade with its North American Neighbours. See generally, Granatstein (1986), p. 11. See also, Crookwell (1990), p. 3.

[62] As will be discussed later, an accession clause for third parties is part of the NAFTA.

[63] As released in a joint statement by the two parties, The White House, Office of the Press Secretary, *Joint Statement by the Presidents of Mexico and the Unite States of a Free Trade Agreement* (undated).

statement on the effects of the proposed agreement.[64] On the subject of labour, trade unionists attacked the agreement on the grounds that NAFTA would cost the United States much needed jobs.[65] The debate on the effects of the agreement on labour in the United States has been contentious, as *The Economist*'s observations of 1992 which reflected:

> NAFTA's effect on employment is the most controversial issue surrounding the agreement, and experts disagree with each other. Figures from the Institute for International Economics show that America stands to lose about 160,000 unskilled factory jobs by 1995, but would also gain about 325,000 from the increased flow of trade. The Economic Policy Institute claims that NAFTA would slow the growth of high-wage jobs by 550,000 in the next 10 years, while depressing American wages.[66]

NAFTA is however, principally designed to achieve an agreement between Canada, the United States and Mexico, which is similar in form and scope to the Canada-United States Free Trade Agreement. The main requirement of this agreement is that it is suitable for North America. Accordingly, NAFTA has integrated policies of the Free Trade Agreement and repealed it. The protagonists of NAFTA recognized the need to have only one agreement, as having two agreements running parallel to each other would have created severe legal difficulties in the field of dispute resolution and rules of origins, resulting in complications in the regulatory structure of North American trade and business. It is with this caution in mind that the drafters of NAFTA culled the principles of the FTA relating to reduction of tariffs, national treatment, government procurement, investment, services and dispute resolution and liberalised them to suit NATA's parties. Another element which formed part of NAFTA was intellectual property rights. Under the Agreement, the Parties also agree unanimously that NAFTA respects the principles of Article XXIV of the GATT relating to the establishment of free trade zones under the most-favoured-nation treatment philosophy.

One of the difficulties faced by the three States Parties to NAFTA was that, unlike in the FTA, there were three Parties involved, making face to face negotiations difficult. Concessions to each Party was made on the basis of how all parties would be affected and what one Party could offer in return to another, for a concession granted. NAFTA changed the overall bilateral concept of the FTA by bringing Mexico into the negotiating room to negotiate on the basis that concessions made by one Party has to benefit the other two.

Canada's decision to officially participate in the negotiations was taken on February 5, 1991, leading to the foundation of NAFTA.[67] This decision was apparently taken by Canada with reluctance, since Canada at that stage viewed its

---

[64] *The Economist*, July 3rd 1993 at 26.

[65] *The Economist*, September 18th 1993 at 27.

[66] *The Economist*, August 1st 1992.

[67] Statement by the Minister for International Trade, Crosbie (1991).

position as being better as a bystander than a participant.[68] However, Canada would have stood to lose a great deal in foreign trade investment opportunities if the United States and Mexico were to sign an exclusive bilateral free trade agreement. Furthermore, the possible exclusion of Canada could have isolated Canada from southern geographic limits applicable to the Agreement, which extends to the tip of Argentina. There was also the strong possibility that the United States would have entered into a series of bilateral free trade agreements with Latin American countries, in the event a bilateral free trade agreement was signed by Mexico and the United States. It was therefore certainly in Canada's best interests to face the challenge of NAFTA and ensure the defence of its fiscal interests.

Mexico's inclusion in NAFTA was the direct result of its change of policy towards international trade and the abandonment of its import substitution policy and its other liberal trade measures. The new policy led to an immediate increase in trade between the United States and Mexico through the burgeoning *Maquiladora* industries. Immediately prior to NAFTA, trade between Mexico and the United States had nearly doubled from 1987 to 1989. At the same time, oil as a Mexican export income earner decreased in trade output from 70 % to 30 %, leading to the need for Mexico to rely more on the United States market. This dependence caused Mexico to need improved access to the American market to create much needed income that had diminished with the gradual decline of its oil revenues. For Mexico, the next logical step towards attracting foreign investment was to stimulate its main export market through a free trade agreement.[69]

NAFTA thus became a vehicle for transporting Mexico's economic revival to the place already occupied by the developed world in international trade. Through NAFTA, Mexico is attempting to foster growth in the manufacturing sector which would in turn create jobs and exports. It also endeavours to raise competitiveness and attract foreign investment. NAFTA's goals are essentially trade-oriented and are calculated to reduce real obstacles to free trade such as burdensome regulations on sanitation, health, safety and subsidies investigations,[70] and obtain equal treatment for investment as a component of an overall free trade zone.

Mexico's role in NAFTA was supported by the successful Framework Trade Agreement between Mexico and the United States, which was signed in 1987.[71] This agreement incorporated a permanent bilateral consultation procedure on issues

[68] See Wilson (1991), p. 4 where it is implied that Canada's participation at the trade talks which led to NAFTA was due to the fact that Canada did not have much choice. See also, Statement of John Weeks, Notes for an Address by John Weeks, Canada's Chief Negotiator for a North American Free Trade Agreement to the Council of the Americas and the Canadian Manufacturers' Association, *Statement no. 91/29*, June 3, 1991 at 3.

[69] Although the Salinas administration expected improved trade with Europe in the early nineties, the unpredicted collapse of the East Bloc of Europe diverted Western Europe's interests in that direction. NAFTA revived the interest of Europe and Japan in Mexico. See Fallows (1991), p. 46.

[70] See Siac (1991). See also, Weintraub (1990), pp. 81–82.

[71] *Framework Trade and Investment Agreement*, reproduced in 1988 27 *International Legal Materials* 438. See also, Smith (1989), p. 655.

such as trade, investment, intellectual property, environment and other issues of interest to both countries. This bilateral alliance was not only successful in increasing trade but was also able to strengthen the Mexico-United States relationship. The consultations related to this agreement laid the foundation for further talks on integration and the formation of other possible treaties between the two countries.

The reason for Mexico's acceptance of Canada as a party to the bilateral alliance with the United States (which resulted in Mexico's acceptance of the NAFTA concept) is because President Salinas was of the view that it was in the interest of both countries to be part of the same agreement with The United States in the long term perspective. The possibility of there being a series of FTAs between the United States and other Latin American countries would have only defeated the objective of communality that NAFTA was seeking in order that foreign investment was encouraged in North America according to a set of unified trade laws.

NAFTA offered for the United States the primary opportunity to secure investments in Mexico and create a separate trading bloc for this purpose. The United States also had cause for optimism in Mexico's trade involvement with the United States and Canada in that a trilateral trade partnership made it easier to obtain more jobs for Americans with access to the Canadian and Mexican markets.[72] Economic growth in the United States was predicted by many studies to result from trade liberalization with Mexico and with the exception of soft sectors which would yield losses, the overall forecast seemed very positive. American exports to Mexico was the pivotal point which created an opening for encouraging trade and investment opportunities for both countries. The driving force behind the perception of the United States was the economic prosperity of Mexico which would indirectly support American economic growth. Mexico had enlarged its market opportunities and indeed, American goods were fashionable in Mexico, thus creating a good opportunity for the United States in Mexico.

A stable economic environment contributed to by the implementation of NAFTA in Mexico will reduce illegal immigration, drug related problems and other security problems in all three countries. A recent commentary explains the benefits that the United States would derive if the migration of Mexicans to the United States was curbed or stopped.[73] NAFTA will also bring for the United States the benefit of forging closer diplomatic ties with Mexico through the implementation of its provisions.

As far as trade relations between the United States and Canada are concerned, NAFTA is considered by the United States Government as an investment that could well iron out some of the rough spots that were caused by the FTA between the two countries.

[72] U.S. Department of Commerce, International Trade Administration, *North American Free Trade Agreement – Generating Jobs for Americans*, U.S. Dept. Comm: Washington, May 1991, at 3.

[73] Weintraub (1990), p. 206.

9.4 Legal Principles Governing NAFTA

9.4.1 The Legal Status of NAFTA at International Law

Treaties, conventions, agreements, protocols, exchanges of notes and other synonyms all mean one and the same thing at international law—that they are international transactions of a legal character. Treaties are concluded between States in written form and governed by international law, whether embodied in a single instrument or in two or more related instruments and whatever its particular designation.[74] Each treaty has four constituent elements: the capacity of the parties thereto to conclude agreement of the provisions of the treaty under international law; the intention of the parties to apply principles of international law when concluding agreement under a treaty; *consensus ad idem* or a meeting of the minds of the parties[75]; and, the parties must have the intention to create legal obligations among themselves. These four elements form a composite regulatory process whereby a treaty becomes strong enough at international law to enable parties to settle their differences within the parameters of the treaty, make inroads into customary international law if necessary, and, transform an unorganized international community into one which may be organized under a uniform set of rules. Treaties are based on three fundamental principles of international law: good faith; consent; and, international responsibility.[76] Since international customary law does not prescribe any particular form for consensual agreements and requirements that would make them binding, the parties to a treaty could agree upon the form of treaty they intend entering into and make it binding among them accordingly. Legal bonds are established between nations because they wish to create them and, as is seen in the *Preamble* to NAFTA, a statement to this effect is

---

[74] *Vienna Convention on the Law of Treaties*, United Nations General Assembly Document *A/CONF.39/27*, 23 May 1969.

[75] There are instances where States may record their reservation on particular provisions of a convention while signing the document as a whole. The International Court of Justice in its examination of the *Genocide Convention* has ruled:

The object and purpose of the Convention…limit both the freedom of making reservations and that of objecting to them. It follows that it is the compatibility of a reservation with the object and purpose of the Convention that must furnish the criterion for the attitude of a State in objecting to the reservation. 1 *I.C.J Rep.* 1951, at 15.

[76] Schwarzenberger and Brown (1976), p. 118.

reflected in the treaty itself.[77] The main feature of a multilateral international agreement is that absolute rights that may have existed within States before the entry into force of such treaty would be transformed into relative rights in the course of a balancing process in which considerations of good faith and reasonableness play a prominent part. However, treaty provisions must be so written and construed as best to conform to accepted principles of international customary law.[78]

Great reliance is placed on treaties as a source of international law. The international Court of Justice, whose function it is to adjudicate upon disputes of an international character between States, applies as a source of law, international conventions which establish rules that are expressly recognized by the States involved in a dispute.[79] The Court also has jurisdiction to interpret a treaty at the request of a State.[80]

The Vienna Convention on the Law of Treaties[81] while recognizing treaties as a source of law, accepts free consent, good faith and the *pacta sunt servanda* as universally recognized elements of a treaty.[82] Article 11 of the Vienna Convention provides that the consent of a State to be bound by a treaty may be expressed by signature, exchange of instruments constituting a treaty, ratification, acceptance,

---

[77] The *Preamble* to NAFTA provides:

The Government of the United States of America, the Government of Canada and the Government of the United Mexican States, resolved to:

STRENGTHEN the special bonds of friendship and cooperation among their nations;
CONTRIBUTE to the harmonious development and expansion of world trade and provide a catalyst to broader international cooperation;
CREATE an expanded and secure market for the goods and services produced in their territories;
REDUCE distortions to trade;
ESTABLISH clear and mutually advantageous rules governing their trade;
ENSURE a predictable commercial framework for business planning and investment;
BUILD on their respective rights and obligations under the General Agreement on Tariffs and Trade and other multilateral and bilateral instruments of cooperation;
ENHANCE the competitiveness of their firms in global markets;
FOSTER creativity and innovation, and promote trade in goods and services that are the subject of intellectual property rights;
CREATE new employment opportunities and improve working conditions and living standards in their respective territories;
UNDERTAKE each of the preceding in a manner consistent with environmental protection and conservation;
PRESERVE their flexibility to safeguard the public welfare;
PROMOTE sustainable development;
STRENGTHEN the development and enforcement of environmental laws and regulations; and,
PROTECT, enhance and enforce basic workers' rights…

[78] Greig (1976), p. 8.

[79] *Statute of the International Court of Justice, Charter of the United Nations and Statute of the International Court of Justice*, United Nations: New York, Article 38.1 (a).

[80] *Id.* Article 36.2 (a).

[81] *Supra* note 74 (this chapter).

[82] *Vienna Convention*, Preamble.

approval or accession, or by any other means agreed upon. "Ratification", "acceptance", "approval", and "accession" generally mean the same thing, *i.e.* that in each case the international act so named indicates that the State performing such act is establishing on the international plane its consent to be bound by a treaty. A State demonstrates its adherence to a treaty by means of the *pacta sunt servanda*, whereby Article 26 of the Vienna Convention reflects that every treaty in force is binding upon the parties and must be performed by them in good faith. The validity of a treaty or of the consent of a State to be bound by a treaty may be impeached only through the application of the Vienna Convention[83] which generally requires that a treaty could be derogated upon only in circumstances the treaty in question so specifies[84]; a later treaty abrogates the treaty in question[85]; there is a breach of the treaty[86]; a *novus actus interveniens* or supervening act which makes the performance of the treaty impossible[87]; and the invocation by a State of the *clausula rebus sic stantibus*[88] wherein a fundamental change of circumstances (when such circumstances constituted an essential basis of the consent of the parties to be bound by the treaty) which has occurred with regard to those existing at the time of the conclusion of the treaty, and which was not foreseen by the parties, radically changes or transforms the extent of obligations of a State. A State may not invoke the fact that its consent to be bound by a treaty has been expressed in violation of a provision of its internal law regarding competence to conclude treaties and seek to invalidate its consent unless such violation was manifest and concerned a rule of its internal law of fundamental importance.[89]

States or international organizations which are parties to such treaties have to apply the treaties they have signed and therefore have to interpret them. Although the conclusion of a treaty is generally governed by international customary law to accord with accepted rules and practices of national constitutional law of the signatory States, the application of treaties are governed by principles of international law. If however, the application or performance of a requirement in an international treaty poses problems to a State, the constitutional law of that State would be applied by courts of that State to settle the problem. Although Article 27 of the Vienna Convention requires States not to invoke provisions of their internal laws as justification for failure to comply with the provisions of a treaty, States are free to choose the means of implementation they see fit according to their traditions and political organization.[90] The overriding rule is that treaties are juristic acts and have to be performed.

---

[83] *Id.* Article 42.1.

[84] *Id.* Article 57.

[85] *Id.* Article 59.

[86] *Id.* Article 60.

[87] *Id.* Article 61.

[88] *Id.* Article 62.1.

[89] *Id.* Article 46.

[90] Reuter (1989), p. 16.

Every international treaty is affected by the fundamental dichotomy where on the one hand, the question arises whether provisions of a treaty are enforceable at law, and on the other, whether the principles of State sovereignty, which is *jus cogens* or mandatory law, would preempt the provisions of a treaty from being considered by States as enforceable. Article 53 of the Vienna Convention addresses this question and provides that where treaties, which at the time of their conclusion, conflict with a peremptory norm of general international law or *jus cogens*, are void. A peremptory norm of general international law is a norm accepted and recognized by the international community of States as a whole as a norm from which no derogation is permitted and which can be modified only by a subsequent norm of general international law having the same character. The use of the words "as a whole" in Article 53 effectively precludes individual States from considering on a subjective basis, particular norms as acceptable to the international community.[91] According to this provision therefore, a treaty such as NAFTA could not have derogated from principles of accepted international legal norms when it was being concluded. The Vienna Convention has, by this provision, implicitly ensured the legal legitimacy of international treaties, and established the principle that treaties are in fact *jus cogens* and therefore are instruments containing provisions, the compliance with which is mandatory.

9.4.2 Rights and Obligations of Parties to NAFTA

The objective of NAFTA is to establish a free trade zone in North America, governed by the three main GATT principles of national treatment, most-favoured-nation treatment and transparency.[92] NAFTA aims at eliminating barriers to trade in, and facilitating the cross border movement of, goods and services between the territories of the parties; promoting conditions of fair competition in the free trade area; increasing substantially investment opportunities in the territories of the parties; providing adequate and effective protection and enforcement of intellectual property rights in each Party's territory; creating effective procedures for the implementation and application of the agreement, for its joint administration and for the resolution of disputes and, establishing a framework for further trilateral, regional and multilateral cooperation to expand and enhance the benefits of the Agreement.[93] The Parties to the agreement are required to apply the provisions of the agreement in the light of the above objectives and in accordance with applicable rules of international law.[94] NAFTA will prevail over any other agreement in the case of any inconsistency between them, except as otherwise provided in

[91] See von der Dunk (1992), pp. 223–224.

[92] For a discussion of these concepts see generally Abeyratne (1994).

[93] NAFTA, Article 102, *International Legal Materials*, Vol XXXII, Number 2, March 1993 at 297.

[94] *Id.* Article 102.2.

NAFTA.[95] NAFTA also provides that each Party shall accord national treatment to the goods of another Party in accordance with Article III of the GATT,[96] including its interpretative notes and to this end Article III of GATT and its interpretative provisions form an integral part of NAFTA.[97] Article 304 of NAFTA precludes any Party from setting arbitrary and self imposed standards and duties pertaining to customs. Furthermore, Article 1103 stipulates that a Party to the agreement shall grant investors of another Party treatment no less favourable than it accords, in like circumstances, to investors of any other Party or of a non Party with respect to the establishment, acquisition, expansion, management, conduct, operation and sale or other disposition of investments.[98] This principle endorses and confirms the most-favoured-nation treatment principle of GATT.

NAFTA is compelling on the subject of environmental issues. Article 1114 (1) provides that Parties are not prevented from acting so as to ensure that investment activity in a territory is undertaken in a manner sensitive to environmental issues. Article 1114(2) explicitly lays down the principle that it is inappropriate for Parties to encourage investment in their territories that would result in relaxing domestic health, safety or environmental health measures. Accordingly, a Party should not waive or otherwise derogate from, or offer to waive or otherwise derogate from such measures as an encouragement for the establishment, acquisition, expansion or retention in its territory of an investment.

As discussed earlier, the issue of environmental protection in relation to the implementation of NAFTA was strongly felt in the United States in particular, where the United States District Court decided that the administration was in breach of the U.S. *National Environmental Policy Act* by not drafting an environmental impact statement on the effects of NAFTA on the U.S. environment.[99] It therefore follows that one of the main concerns in the implementation of NAFTA is to protect the environment of the territory in which investments are made under the Agreement, that are calculated to endanger or in any manner harm the environment therein.

---

[95] *Id.* Article 103.2.

[96] Article III of GATT stipulates that the contracting Parties to GATT recognize that internal taxes and other internal charges and laws, regulations and requirements affecting the internal sale, offering for sale, purchase, transportation, distribution or use of products, and internal quantitative regulations requiring the mixture, processing or use of products in specified amounts or proportions, should not be applied to imported or domestic products so as not to afford protection to domestic production. Furthermore, the provision also states that the products of a territory of any contracting Party imported into the territory of any other contracting Party shall not subject, directly or indirectly, to any taxes or other internal charges of any kind in excess of those applied, directly or indirectly, to like domestic products. Moreover, no contracting Party shall otherwise apply internal taxes or other internal charges to imported or domestic products in a manner contrary to the principles set forth in the Agreement. GATT, *supra*, Chap. 5, note 21.

[97] *Id.* Article 301.

[98] See Article 1139 of NAFTA for a definition of "investment".

[99] *The Economist*, July 3rd 1993 at 26.

By Article 1102 (1) and (2), NAFTA provides that each Party shall accord to investors of another Party treatment no less favourable than that it accords, in like circumstances, to its own investors and investments of investors with respect to the establishment, acquisition, expansion, management, conduct, operation and sale or other disposition of investments. These provisions are also extended to apply to service providers of each Party.[100] There is also a minimum standard of treatment prescribed in Article 1105 for investments of investors, such standard to accord with principles of international law, ensuring *inter alia*, fair and equitable treatment and full protection and security.

Another significant provision in the Agreement is Article 1406 which requires each Party to accord to investors of another Party, financial institutions of another Party, investments of investors in financial institutions and cross-border financial service providers of another Party treatment no less favourable than that it accords to the investors, financial institutions, investments of investors in financial institutions and cross-border financial service providers of any other Party or of a non Party in like circumstances.

NAFTA therefore creates responsibility on the part of a Party to the Agreement towards investments, investors and financial institutions of other Parties. Professor Brownlie remarks:

> Today, one can regard responsibility as a general principle of international law, a concomitant of substantive rules and of the supposition that acts and omissions may be categorized as illegal by reference to the rules establishing rights and duties. Shortly, the law of responsibility is concerned with the incidence and consequences of illegal acts, and particularly the payment of compensation for loss caused.[101]

Thus, a breach of any of the provisions of NAFTA, resulting in financial loss to investors and financial institutions would, on the basis of the national treatment and most-favoured-nation-treatment principle, necessarily entail the payment of compensation to the aggrieved Party by the Party which breaches such provisions. Professor Brownlie also contends that the nature of State responsibility is not based upon delict in the municipal sense, and international responsibility relates both to breaches of treaty and to other breaches of legal duty, extending responsibility of States to individuals as well as other States. This view is supported by Justice Huber in the *Spanish Zone of Morocco Claims* case[102]—that international rights involve international responsibility and a dereliction of such responsibility would necessarily entail making of reparation by the Party causing such breach.[103] The Provisional International Court of Justice (PCIJ) remarked earlier:

> It is a principle of international law that the breach of an engagement involves an obligation to make reparation in an adequate form. Reparation therefore is the indispensable

---

[100] Articles 1202, 1203 and 1204.

[101] Brownlie (1990), p. 433.

[102] *RIAA* ii 615 at 641.

[103] *Ibid.*

complement of a failure to apply a Convention and there is no necessity for this to be stated in the Convention itself[104]

Any investment, therefore, which is endangered and subsequently lost to the investor owing to the breach of a provision by a Party to the Agreement would be reimbursable by that Party. Since the Agreement itself has provided for the protection of investments, anyone of the three State Parties to NAFTA may also invoke the principles of public international law to protect their investments, or those of their nationals.

NAFTA therefore creates duties upon its Parties in two areas:

a) protection and security by one Party of investments of investors of other Parties to the Agreement according to the principles of international law, and fair and equitable treatment;
b) treatment by one Party, of investors, their investments, financial institutions and cross-border financial service providers of other Parties to the Agreement on a basis no less favourable than accorded to others, whether they are Parties to the Agreement or not.

These two obligations explicitly lay down the duties of one Party to the Agreement as being general in regard to investments and specific in regard to investors and financial institutions of other Parties. In other words, a Party to the Agreement has to specifically abide by the general principles of international law in its treatment of investments of investors of another Party, while it has only to follow the national and most-favoured-nation treatment approach in regard to investors and financial institutions of another Party.

With the signing of NAFTA by Canada, Mexico and the United States, free trade between the three countries fell into place. The primary concerns of the United States in the implementation of the Agreement were related to environmental and labour issues as well as the fear that possible "import surges" from Mexico could wipe out American businesses. Canada fears that free trade under NAFTA would suck jobs south. Mexico on the other hand, while standing to gain from increased investment in Mexico, is ever watchful of the trade relations that Canada and the United States might enter into under the Agreement. As far as the Canadian and the American apprehensions are concerned, it is now felt that the number of jobs lost from the implementation of NAFTA is likely to be small—about half a million over the next decade. Considering that the United States displaced 20 million jobs in the 1980s, the effect of NAFTA is expected to result in a small net increase in employment in the United States. The Agreement in general is perceived as portending more good than evil for the three countries concerned and it is this reason that has led to its being signed in December 1992 by them without further debate.

[104] *(1927) PCIJ Ser. A. No. 17* at 29.

References

Abeyratne RIR (1994) The liberalization of air transport services within GATT – some legal issues. Trading Law Trading Law Rep 13(1):2

Albrecht J (1984) Canadian foreign investment policy and the international politico legal process. Canadian Yearbook of International Law 1982. University of British Columbia Press, Vancouver, p 149

Arnett EJ (1985) From FIRA to investment Canada. Alberta Law Rev 24:1

Baker J (1984) From FIRA to investment Canada. In: Fry E, Radebaugh L (eds) Canada-U.S. economic relations in the conservative era of Mulroney and Regan. Brigham University, Provo, Utah, p 47

Baker MB (1988) Debt-equity swaps and Mexican law: the interplay between law and regulation. North West J Int Law 9:333

Bale HE Jr (1985) The United States policy towards inward foreign direct investment. Vand J Trans Law 18:199

Brownlie I (1990) Principles of public international law, 4th edn. Clarendon Press, Oxford, p 433

Castel JG (1989) Extraterritoriality in international trade. Butterworth, Toronto

Crookwell H (1990) Canadian American trade and investment under the free trade agreement. Quorum Books, New York, p 3

Crosbie J (1991) Canada-U.S.-Mexico free trade negotiations, Press Release 91/09. External Affairs and International Trade, Canada, February 5

Dallemeyer DG (1989) Foreign policy and export controls: how will the Canada-United States free trade agreement accommodate the extra-territorial application of United States laws to Canadian exports of goods and technology? Georgia J Int Comp Law 19:565

Davidow J (1976) Extraterritorial application of U.S. antitrust law in a changing world. Law Policy Int Bus 8:895

Dewhirst G, Rudiak M (1986) From investment screening to investment development: the impact of Canada's Foreign Investment Review Agency (FIRA) and investment Canada in Canada's technological development. Can US Law J II:149

Dewhirst M (1984) The Canadian federal government's policy in foreign direct investment. In: Fry E, Radebaugh L (eds) Regulation of foreign direct investments in Canada and the United States. Brigham University, Provo, p 27

du Mars TC, Beltran el Rio S (1988) A survey of the air and water quality laws of Mexico. Nat Resour J 28:787

Fallows J (1991) The romance with Mexico. The New York Review, November 7, p 46

Frank TF, Gudgeon KS (1975) Canada's foreign investment control experiment: the law, the context and the practice. N Y Univ Law Rev 50:76

Glover GC, New DC, Lacourciere MM (1985) The Investment Canada Act: a new approach to foreign investments in Canada. Bus Lawyer 4:83

Graham EM, Krugman PR (1989) Foreign direct investment in the United States. Institute for International Economics, Washington, p 127

Granatstein JL (1986) L'Éternelle question du libre-échange de développement du Canada. Les dimensions politiques des rapports économiques entre le Canada et les États-Unis, vol 29. Ministère des Approvisionnements et Services, Ottawa, p 11

Greig DW (1976) International law, 2nd edn. Butterworths, London, p 8

Heinz Juergen M (1986) Regulation of foreign investment in Canada. LL.M Thesis, McGill University, Canada

Hodgson JH (1975) Las Inversiones Estranjeras en el desarollo de Canada y de America Latina. In: Farell RB (ed) America Latina y Canada Frente a la Politica Exterior de los Estados Unidos. Fondo de Cultura Economica, Mexico, D.F., p 56

Kate AT (1978) La politica de protection en el desarollo de Mexico. Fondo de Cultura Economica, Mexico

Knee JA (1989) Limiting abuse of Exon-Florio by takeover targets. George Wash J Int Law Econ 23:475

Kraft JA (1988) Recent development, antitrust law: extra-territoriality: In re: Uranium Anti Trust Litigation. Harv J Int Law 21:515

MacDonald HI (1972) Nationalism in Canada. In: Reid TE (ed) Foreign ownership: Villain or Scapegoat. Holt, Reinhart and Winston, Toronto, p 71

McCambly J (1991) North American free trade requires a charter of labour. Can Speeches/Issues 4:47

McDowall D (1984) A fit place for investment? Study No. 81. Conference Board of Canada, Ottawa

Mendenhall J (1991) Recent developments: U.S. executive authority to divest acquisitions under the Exon-Florio amendment – the MAMCO divestiture. Harv J Int Law 32:285

Mendes E (1981) The Canadian energy program: an example of assertion of economic sovereignty or creeping expropriation in international law. Vand J Trans 475

Mumme SP et al (1988) Political development and environmental policy in Mexico. Latin Am Res Rev 23:7

Nance DS, Wasserman J (1990) Regulation of imports and foreign investment in the United States on national security grounds. Mich J Int Law 11:926

Olmstead CJ, Krauland EJ, Orentlicher DF (1984) Expropriation in the energy industry: Canada's crown share provisions as a violation of international law. McGill Law J 439

Prichard M (1988) Status of the omnibus trade bill and the Canada-U.S. free trade agreement. Rev Int Bus Law 2:95

Raby J (1990) The investment provisions of the Canada-United States free trade agreement: a Canadian perspective. Am J Int Law 395

Reuter P (1989) Introduction to the law of treaties. Pinter Publishers, London and New York, p 16

Reyes ST (1991) Labour market interdependence between Mexico and the United States. In: Reynolds CW (ed) The dynamics of North American trade and investment – Canada, United States, Mexico. Stanford University Press, Stanford, p 241

Sandstrom M, Coccuza C (1989) The Omnibus Trade and Competitiveness Act of 1988: an overview. Rev Int Bus Law 3:65

Scarborough P (1985) The foreign investor in the United States: disclosure, taxation and visa laws. Int Lawyer 19:85

Schwarzenberger G, Brown ED (1976) A manual of international law, 6th edn. Professional Books Limited, Oxon, p 118

Siac RC (1991) The use and abuse of unfair trade remedy laws: the Mexican U.S. experience – implications for trilateral free trade negotiations. C.D. Howe Institute, Toronto, January

Smith G (1989) The U.S.-Mexico framework agreement: implications for bilateral trade. Law Policy Int Bus 20:655

Spence JM (1986) Current approaches to foreign investment review in Canada. McGill Law J 31:507

Swan PN (1984) International antitrust—the reach and efficacy of United States law. Oreg Law Rev 63:177

Torres Landa JF (1990) Report on the new rules for operation of debt-equity swaps in Mexico. Int Lawyer 25:733

Turner J (1983) Canadian regulation of foreign direct investments. Harv Int Law J 23:333

von der Dunk FG (1992) Jus Cogens Sive Lex Ferenda: Jus Cogendum. In: Masson-Zwaan TL, Mendes De Leon PMJ (eds) Air and Space Law: De Lege Ferenda, Essays in Honour of Henri A Wassenbergh. Martinus Nijhoff, Dordrecht, p 219

Weintraub S (1990) A marriage of convenience – relations between Mexico and the United States. Oxford University Press, New York

Wex S (1984) Instead of FIRA: autonomy for Canadian subsidiaries? Institute for Research on Public Policy, Montreal

Wilson MH (1991) Notes for a speech by the honourable M./H. Wilson at the financial post conference on North American free trade. International Trade Statement, Np. 91/22, April 25, p 4

Chapter 10
Competition Law in Air Transport

10.1 Introduction

Airlines have to change and adapt to the world economy.[1] At the time of writing, the economies of the developed countries of Europe and North America were slowing down, at least in terms of growth compared to pre-financial crisis times of 2008–2010 and there was rapid economic growth in the East in the emerging economies. Concurrently, deep and contentious debates were taking place among regulators on aviation and climate change that were focused on air transport in particular.[2] Against this backdrop, the buzz word was *connectivity*, on which many airlines anchored their competitive stance, much to the annoyance of the proponents of protectionism. The air transport industry (*i.e.* the airlines) like any other industry has to formulate its competitive strategy on the conduct of certain inevitable forces, such as new entrants into the market; disruptive innovation of competitors; customers and clients' changing demands; bargaining power of suppliers; and demands of the workforce. Cost reduction and product variance, as needed, are the main drivers that cope with the forces of competition. Added to these key determinants are economies of scale, where large scale production bring costs down, and capital investment which helps enhance quantitative and qualitative production. These form a cocktail of stimuli for robust competition, provided government regulation is conducive to competition in air transport.

The Organization for Economic Cooperation and Development (OECD) identifies with having always a product and a geographic dimension. The OECD strictly relies on origin/destination as the determinable market in the context of air transport and the transportation of persons and goods from one point to the other. Consequently, the delineation of relevant markets in the airline sector typically starts with

[1] See generally, Abeyratne (2012a).

[2] See generally, Abeyratne (2013a).

R. Abeyratne, *Competition and Investment in Air Transport*,
DOI 10.1007/978-3-319-24372-6_10

the identification of the point of origin (O) and the point of destination (D), the so-called O&D approach.[3]

In terms of government regulation, the hapless airlines have been faced with ambivalent and often convoluted policy or non-policy, spearheaded by an equally confounded ICAO which has so far come up with only one statement—that ICAO endorses liberalization of air transport, which should be pursued in accordance with the particular circumstances of each member State.[4] This has led to a dichotomy that revolves round the competitiveness of nations and the role of government in industrial growth and their contribution to robust competition. On the one hand, proponents of governmental control argue that the government should help and support industry, while the other argument is that government should wash its hands off industry and give it a free hand. Michael Porter says both views are incorrect:

> Government's proper role is as a catalyst and challenger: it is to encourage – or even push – companies to raise their aspirations and move to higher levels of competitive performance, even though this process may be inherently unpleasant and difficult. Governments cannot create competitive industries, only companies can do that.[5]

This view gives some credence to the policy of Emirates Airlines—a much maligned carrier by the protectionist lobby which accuses the airline of stealing market share by using government subsidies. Emirates says that at the core of its business model is "a commitment to true international competition and open skies". The philosophy of the airline is based on the belief that "an open global economy is vital to free and fair trade, economic growth and fuller employment". Emirates Airlines goes on to say:

> The protection of a national carrier simply because it carries the flag of one particular country belongs to the past. Where governments protect flag carriers, competition is weak and these routes often have some of the highest fares in the world. We believe that market access is vital for exporters, passengers, airports and local economies. Robust competition lowers prices and allows more people to fly, and liberalised economies with open market access tend to show the strongest performance. Dubai is a long-term supporter of open skies with over 150 international airlines flying to its airport.[6]

The aviation industry of the United Arab Emirates (UAE) provides nearly $40 billion to the economy of the UAE. Contrary to the claim that the airlines in the Emirates survive on handouts from the governments it seems to be the other way around, where governments, as Porter says, provide the stimulus and act as catalysts to the industry. Tony Tyler, Director General of the International Air Transport Association (IATA) said at the World Passenger Symposium held in Abu Dhabi in

[3] AIRLINE COMPETITION – Background Paper by the Secretariat, DAF/COMP(2014)14
Organisation de Coopération et de Développement Économiques, Organisation for Economic Co-operation and Development, 12-Jun-2014, at 22.

[4] See generally, Abeyratne (2012b). Also see Abeyratne (2013b).

[5] Porter (1998), p. 184.

[6] http://www.emirates.com/english/about/int-and-gov-affairs/international-and-government-affairs-new.aspx.

2015 that governments in the UAE understand that connectivity in air transport can be key drivers of their economies and in pursuance of this approach they created a business-friendly environment for air transport with low taxes and superior infrastructure.[7]

A similar statement by James Hogan, CEO of Etihad Airways—the carrier based in Abu Dhabi, while recognizing that the airline is owned by the government of Abu Dhabi stated at the US Chamber of Commerce Foundation's 14th Annual Aviation Summit in Washington, DC in March 2015, that Etihad received equity investment and shareholder loans, which have been supplemented by US$10.5 billion in loans from international financial institutions. The basis for these investments was that the government believed in the airline's business plan and certain success in the airline.[8] Hogan pointed out that the competitive advantage gained by Etihad—an airline which started with a blank sheet from scratch—was due to nothing more than incredible customer service, delivered on modern new aircraft, with a world-leading product, at competitive prices, on routes people want to fly.[9]

QATAR Airways—the third maligned Gulf carrier—has a similar tale to tell. Abdul Aziz Al Noaimi, Chairman and Chief Executive Officer of the Civil Aviation Administration of Qatar, in an interview given in 2015 categorically asserted that airlines should be able to compete with each other like in any other business and that restrictive bilateral air services were a negative influence in the industry.[10] This brings the discussion to competition laws in air transport.

10.2 Competition Laws in Air Transport

The United Nations General Assembly in December 1980 adopted Resolution A 35/63 which in turn adopted a multilateral set of rules calculated to control restrictive business practices. Later, at its 83rd plenary Meeting in December 2006 The Assembly adopted Resolution 61/186 (International Trade and Development), which reaffirmed *inter alia* that the fundamental role that competition law and policy can play for sound economic development and the validity of the Set of Multilaterally Agreed Equitable Principles and Rules for the Control of Restrictive Business Practices, as well as the important and useful role that the United Nations

[7] Who Controls the World? *The National*, Business Segment, http://www.thenational.ae/business/industry-insights/aviation/aviation-provides-big-lift-for-uae-economy.

[8] https://www.etihad.com/en-us/about-us/etihad-news/archive/2015/open-skies-is-a-model-of-success-says-etihad-airways-president-and-chief-executive-officer/.

[9] *Ibid.*

[10] https://books.google.ca/books?id=BT1gIeuQq6wC&pg=PA125&lpg=PA125&dq=government+as+a+catalyst+for+Qatar+Airways&source=bl&ots=IjtXoW4pmv&sig=QMeqh9rbFPJIiAn3lzjII4nM-IA&hl=en&sa=X&ei=4ZWaVYWxCdKhyATY3YKgDA&ved=0CEoQ6AEwBw#v=onepage&q=government%20as%20a%20catalyst%20for%20Qatar%20Airways&f=false.

Conference on Trade and Development plays in this field, and decided to convene in 2010, under the auspices of the United Nations Conference on Trade and Development (UNCTAD). Section B of the UN Principles defines restrictive business practices as acts or behaviour of enterprises which, through an abuse or acquisition and abuse of a dominant position of market power, limit access to markets or otherwise unduly restrain competition, having or being likely to have adverse effects on international trade, particularly that of developing countries, and on the economic development of these countries, or which through formal, informal, written or unwritten agreements or arrangements among enterprise have the same impact.

In 2004 UNCTAD developed a *Model Law on Competition*, the purpose of which is to control or eliminate restrictive agreements or arrangements among enterprises, or mergers and acquisitions or abuse of dominant positions of market power, which limit access to markets or otherwise unduly restrain competition, adversely affecting domestic or international trade or economic development. The Model Law prohibited agreements between rival or potentially rival firms, regardless of whether such agreements are written or oral, formal or informal if they were: agreements fixing prices or other terms of sale, including in international trade; collusive tendering; market or customer allocation; restraints on production or sale, including by quota; concerted refusals to purchase; concerted refusal to supply; collective denial of access to an arrangement, or association, which is crucial to competition.

Various countries have their own legislation and interpretation of anti competitive practices and dominant positions of competitors. To give just a few examples, in *China*, according to article 17 of the Anti-Monopoly Law of China, a dominant market position is defined as a market position held by a business operator that has the ability to control the price or quantity of commodities or other trading conditions in the relevant market or to hinder or affect the entry of other business operators into the relevant market. Furthermore, six main factors to determine a dominant market position of a business operator are provided under article 18: the market share of the business operator and its competitive status in the relevant market; the ability of the business operator to control the sales market or the raw material supply market; the financial and technological conditions of the business operator; The extent of reliance on the business operator by other business operators in the transactions; the degree of difficulty for other business operators to enter the relevant market; and other factors relevant to the determination of the dominant market position of the business operator.

In *South Africa*, The Competition Act, 1998 (No. 89), article 7, establishes that a firm is dominant in a market under the following conditions: (a) it has at least 45 % of that market; (b) it has at least 35 %, but less than 45 % of that market, unless it can show that it does not have market power; or (c) it has less than 35 % of that market, but has market power. In *India*, The Indian Competition Act, 2002 defines dominant position under article 4 as a "position of strength, enjoyed by an enterprise, in the relevant market, in India", which enables it to "operate independently of competitive forces prevailing in the relevant market or affect its competitors or consumers or the relevant market in its favour". The Competition Commission of India, when inquiring whether an enterprise enjoys a dominant position or not, has due regard to all or any of these factors. The same article states that no enterprise or group shall abuse its dominant position. There shall be an abuse of dominant position if an enterprise directly or indirectly imposes unfair or discriminatory conditions on the purchase or sale of goods or services or unfair or discriminatory prices, including predatory prices, on the purchase or sale of goods or services.

In *Estonia*, dominance requires that an undertaking be able to operate to an appreciable extent independently of competitors, suppliers and buyers. Dominance is presumed if an undertaking or several undertakings hold a market share of more than 40 % of the turnover in the market. Undertakings with special or exclusive rights, or in control of essential facilities, are also considered as dominant; see paragraph 13 of the Competition Act of 2001.

In *Zambia*, The Competition and Consumer Protection Act of 2010 (No. 24), part III, article 15, indicates that a dominant position exists in relation to the supply of goods or services if (a) thirty per cent or more of those goods or services are supplied or acquired by one enterprise; or (b) sixty per cent or more of those goods or services are supplied or acquired by not more than three enterprises. Article 16 establishes the prohibition of abuse of dominant position stating that an enterprise shall refrain from any act or conduct if, through abuse or acquisition of a dominant position of market power, the act or conduct limits access to markets or otherwise unduly restrains competition, or has or is likely to have adverse effect on trade or the economy in general.

In the *Czech Republic* Article 10 (1) of the Consolidated Act on the Protection of Competition (2001) states that one or more undertakings jointly (joint dominance) shall be deemed to have a dominant position in the relevant market if their market power enables them to behave independently, to a significant extent, of other undertakings or consumers. According to article 10 (3), unless proven otherwise, an undertaking or undertakings in joint dominance is deemed not to be in a dominant position if its/their share of the relevant market achieved during the period examined does not exceed 40 %.

In *Germany*, according to the Act against Restraints of Competition, paragraph 19, an undertaking is dominant where, as a supplier or purchaser of certain kinds of goods or commercial services in the relevant product and geographic market, it has no competitors or is not exposed to any substantial competition, or it has a paramount market position in relation to its competitors. For this purpose, account shall be taken in particular of its market share, financial power, access to supplies or markets, links with other undertakings, legal or factual barriers to market entry by other undertakings, actual or potential competition by undertakings established within or outside the scope of application of the Act, and its ability to shift its supply or demand to other goods or commercial services, as well as the ability of the opposite market side to resort to other undertakings. Two or more undertakings are dominant insofar as no substantial competition exists between them with respect to certain kinds of goods or commercial services and they jointly satisfy the conditions set out above. An undertaking is presumed to be dominant if it has a market share of at least one third. A number of undertakings are presumed to be dominant if (a) they consist of three or fewer undertakings reaching a combined market share of 50 % or (b) they consist of five or fewer undertakings reaching a combined market share of two thirds, unless the undertakings demonstrate that the conditions of competition may be expected to maintain substantial competition between them, or that the number of undertakings have no paramount market position in relation to the remaining competitors reaching a combined market share.

10.2.1 State Aid

It would not be incorrect to say that the genesis of competition law is consumer welfare and protection. At the 38th Session of ICAO Assembly in 2013 the Assembly adopted a Resolution which called upon the Council of ICAO to Request the Council to develop, in the short term, a set of high-level, non-binding, non-prescriptive core principles on consumer protection, for use as policy guidance, which strike an appropriate balance between protection of consumers and industry competitiveness and which take into account the needs of States for flexibility, given different State social, political and economic characteristics. Regrettably, these non-binding, non-prescriptive consumer protection principles would be as impotent as any other ICAO documents offering States "guidance". Presumably, the Assembly was not aware, nor did care that there are established laws under common law, particularly applicable in the European Union States and the United Sates that prohibit anti competitive conduct, which have been established through a *cursus curiae* that clearly spells out how an economic entity should conduct business.

The European Commission (EC), in a clear statement made in 2005 drew the intrinsic link between competition and consumer protection and recognized that consumers can be harmed by distortions in the competitive structure of the market. The Statement of the EC said that consumer welfare, being an established principle in the Commission and that competition must be protected so that consumer interests are protected and resources are properly allocated.[11] This means that economic entities have to compete at the same level, offering the consumer a choice of different product at a reasonable price.

At competition law, airlines are undertakings which engage in an economic activity irrespective of their constructs and financial establishment. Any activity offering goods and services in a given market is an economic activity. Under European law,[12] the basic instrument that governs competition is the Treaty on

[11] Speech of 15 September 2005 at www.ec/europa/eu/competition.

[12] The European Community was established in 1957 by the Treaty of Rome and initially comprised six Member States. Following the entry into force of the Treaty of Lisbon on 1 December 2009, the EC Treaty is called the Treaty on the Functioning of the European Union

the Functioning of the European Union (TFEU) 2012/C 326/0, Articles 107 to 109 address the question of State aids. Article 107 (1) provides that any aid granted by a Member State or through State resources in any form whatsoever which distorts or threatens to distort competition by favoring certain undertakings or the production of certain goods shall, in so far as it affects trade between Member States, be incompatible with the internal market. Article 108 prescribes that The Commission shall, in cooperation with Member States, keep under constant review all systems of aid existing in those States. It is required to propose to the latter any appropriate measures required by the progressive development or by the functioning of the internal market. If, after giving notice to the parties concerned to submit their comments, the Commission finds that aid granted by a State or through State resources is not compatible with the internal market having regard to Article 107, or that such aid is being misused, it has to decide that the State concerned shall abolish or alter such aid within a period of time to be determined by the Commission.

If the State concerned does not comply with this decision within the prescribed time, the Commission or any other interested State may refer the matter to the Court of Justice of the European Union direct. On application by a Member State, the Council may, acting unanimously, decide that aid which that State is granting or intends to grant shall be considered to be compatible with the internal market, in derogation from the provisions of Article 107 or from the regulations provided for in Article 109, if such a decision is justified by exceptional circumstances. If, as regards the aid in question, the Commission has already initiated the procedure provided for in the provision, the fact that the State concerned has made its application to the Council would have the effect of suspending that procedure until the Council has made its attitude known. If, however, the Council has not made its attitude known within 3 months of the said application being made, the Commission is required to give its decision on the case.

Article 108 further provides that the Commission needs to be informed, in sufficient time to enable it to submit its comments, of any plans to grant or alter aid. If it considers that any such plan is not compatible with the internal market having regard to Article 107, it has to, without delay initiate the procedure provided for in the Article. The Member State concerned cannot put its proposed measures into effect until this procedure has resulted in a final decision. The Commission may adopt regulations relating to the categories of State aid that the Council has, pursuant to Article 109, determined may be exempted from the procedure provided for by paragraph 3 of Article 108.

Article 109 stipulates that The Council of Europe, on a proposal from the Commission and after consulting the European Parliament, may make any appropriate regulations for the application of Articles 107 and 108 and may in particular

(TFEU). The European Union comprises 27 Member States: Austria, Belgium, Bulgaria, Cyprus (Greek), the Czech Republic, Denmark, Estonia, Finland, France, Germany, Greece, Hungary, Ireland, Italy, Latvia, Lithuania, Luxembourg, Malta, the Netherlands, Poland, Portugal, Romania, the Slovak Republic, Slovenia, Spain, Sweden, and the United Kingdom.

determine the conditions in which Article 108(3) applies and the categories of aid exempted from the procedure.

Where the external market is concerned, as would apply to the operations of the Gulf carriers into and out of Europe, Regulation (EC) No 868/2004 of the European Parliament and of the Council of 21 April 2004 concerning protection against subsidization and unfair pricing practices causing injury to Community air carriers in the supply of air services from countries not members of the European Community applies. According to the regulation, a subsidy exists when a government, regional body or other public organisation makes a financial contribution that confers a benefit. It may take the form of: grants, loans or equity infusion, potential direct transfer of funds or the assumption of liabilities; revenue that is otherwise due but which is foregone or not collected; the supply of goods or services other than general infrastructure, or their purchase by a public body; payments by a public body to a funding mechanism or the entrusting to a private body of one of the functions described above.

If there is a breach of Regulation 868/2004 a complaint can be initiated and an investigation is commenced when a written complaint is lodged by the EU industry or on the Commission's own initiative. Where sufficient evidence exists, the proceeding is initiated within 45 days of the lodging of the complaint, but this period may be extended by up to 30 days if there is a bilateral agreement. Notice of the initiation of the procedure must be published in the Official Journal and include the details specified in the regulation. The Commission must notify interested parties. The Commission has 45 days within which to inform the complainant if insufficient evidence is presented.

This investigation is required to be concluded within 9 months of proceedings being initiated. An extension may be allowed if a satisfactory resolution of the complaint appears imminent or if additional time is needed in order to achieve a resolution that is in the EU interest. Interested parties may be granted a hearing. However, if they refuse access to or fail to provide necessary information within the appropriate time limits, the final findings may be made on the basis of facts available.

Four possible scenarios may be the result of an investigation: provisional measures—these may be imposed for a maximum period of 6 months if it is determined that injury is being caused and that the EU interest calls for intervention to prevent further such injury; termination of the proceedings without measures being imposed—this happens when the complaint is withdrawn or a satisfactory remedy is obtained; definitive measures: these are imposed when it is established that unfair pricing practices or subsidies which cause injury exist. The level of measures imposed must not exceed the level of the subsidies or the difference between the fares charged by the two air carriers concerned (EU and non-EU)—and undertakings: an investigation may be terminated without measures being imposed if the public authorities or non-EU air carrier concerned undertake to eliminate the subsidies and revise its prices in order to prevent further injury. In the event of an undertaking being breached, a definitive measure will be imposed. Should the circumstances warrant, the Commission may review the imposition of the measures in their initial form with a view to repealing, modifying or maintaining them.

It must be noted that in the European Union, a member State is generally not allowed to give aid to a weakening or ailing company. This is based on the reasoning that other competitors may be unduly disadvantaged by a bailout to the ailing company and that State aid would be unfair by the consumer who would be deprived of the various choices offered by the competitors, if they are forced to pull out of the market. On the face of it this philosophy seemingly makes no sense unless the EU bases its argument on the theory of creative destruction where an ailing company would die if it cannot survive a financial crisis. Some have argued that such a bailout would be beneficial to the State economy concerned.[13] However, this general principle has been obviated in the face of companies facing difficulties and the EU has sanctioned State aid in several circumstances in the case of air carriers.

In the 1993 case[14] involving the Belgian career SABENA The European Commission allowed the Belgian government to give restructuring aid to the ailing airline. Conditions were attached to the bailout however, to the extent that the government would not get involved in the management of the airline and that there would be no further injections of cash. *Truxal* discusses many instances of European State aid, including aid granted to *TAP Air Portugal; Iberia; Olympic Airlines; Air Lingus; Austrian Airlines; Ryanair; Cyprus Airways* and *Maliev.*[15]

10.2.2 Dominant Position

Article 82 of The Treaty Establishing the European Community (Treaty of Rome of 27 March 1957) prohibits abuses of a dominant position. In accordance with the case-law, it is not in itself illegal for an undertaking to be in a dominant position and such a dominant undertaking is entitled to compete on the merits. However, the undertaking concerned has a special responsibility not to allow its conduct to impair genuine undistorted competition in the common market. Article 82 is the legal basis for a crucial component of competition policy and its effective enforcement helps markets to work better for the benefit of businesses and consumers. This is particularly important in the context of the wider objective of achieving an integrated internal market.

The Treaty, in Article 86 provides that any abuse by one or more undertakings of a dominant position within the common market or in a substantial part of it shall be prohibited as incompatible with the common market in so far as it may affect trade between Member States. Such abuse may, in particular, consist in: directly or indirectly imposing unfair purchase or selling prices or unfair trading conditions;

[13] See Anderson (2004), p. 544.

[14] (1991) OJ L 300/48, 31.10.2001.

[15] See Truxal (2012), pp. 93–102.

limiting production, markets or technical development to the prejudice of consumers; applying dissimilar conditions to equivalent transactions with other trading parties, thereby placing them at a competitive disadvantage; or making the conclusion of contracts subject to acceptance by the other parties of supplementary obligations which, by their nature or according to commercial usage, have no connection with the subject of such contracts.

Council Regulation (EC) No 1/2003 of 16 December 2002 on the implementation of the rules on competition laid down in Articles 81 and 82 of the Treaty adopted by the Council on 16 December 2002 which implements the rules on competition laid down by Articles 101 and 102 of the Treaty on the Functioning of the European Union (TFEU) (formerly Articles 81 and 82 of the Treaty establishing the European Community (EC Treaty)), replaced Regulation EEC No 17/62 from 1 May 2004.

Regulation No 17/62 established a centralised monitoring system under which agreements liable to restrict and affect trade between EU countries must, in order to qualify for an exemption, be notified to the Commission. The Commission's exclusive power to authorise agreements which restrict competition but which meet the conditions of Article 81(3) of the EC Treaty has led to a large number of agreements being notified by companies, a fact which has undermined efforts to promote a rigorous and decentralised application of the EU competition rules.

The TFEU which came later, prescribed in Articles 101 and 102 the principles of dominant position as applicable in the EU. Article 101 provides that prohibits as incompatible with EU principles, all agreements between undertakings, decisions by associations of undertakings, and concerted practices which may affect trade between Member States and which have as their object or effect the prevention, restriction, or distortion of competition within the EU. The following must be established for an infringement of Article 101(1): an agreement or concerted practice between two or more undertakings, or a decision by an association of undertakings; which has as its object or effect the prevention; restriction or distortion of competition; and an appreciable effect on competition; and an appreciable effect on trade between Member States.

Article 102 in particular has been adopted with a view to preventing undertakings who hold a dominant position in a market from abusing that position. Its fundamental purpose and role is to regulate monopolies, which works to restrain and limit competition in private industry and produce worse outcomes for consumers and society. Article 102 provides that any abuse by one or more undertakings of a dominant position within the internal market or in a substantial part of it shall be prohibited as incompatible with the internal market in so far as it may affect trade between Member States. Such abuse may, in particular, consist in: directly or indirectly imposing unfair purchase or selling prices or other unfair trading conditions; limiting production, markets or technical development to the prejudice of consumers; applying dissimilar conditions to equivalent transactions with other trading parties, thereby placing them at a competitive disadvantage; making the conclusion of contracts subject to acceptance by the other parties of supplementary obligations which, by their nature or according to commercial usage, have no connection with the subject of such contracts. One commentator says that Article 3 (1)g of the Treaty of European Communities recognized the vital importance of establishing a system ensuring that competition in the internal market is not distorted.[16]

The European Court of Justice, in a 1978 decision pronounced that the dominant position principle as enunciated by Article 102 should be interpreted as a position of economic strength created by an undertaking which enables it to effectively preclude competition, giving in turn that undertaking economic independence over its competitors, customers and consumers.[17] The term "dominant position" is not explicitly defined anywhere and is deemed to mean substantial market power.

10.2.3 Competition Laws in the United States

Ex facie, the competition laws that prevail in Europe are similar to those in the United States. For example, Article 85 of the Treaty of Rome which provides against market distortion is similar to Section 1 of the Sherman Antitrust Act of 1890 which States that every contract, combination in the form of trust or otherwise, or conspiracy, in restraint of trade or commerce among the several States, or with foreign nations, is declared to be illegal. Every person who shall make any contract or engage in any combination or conspiracy hereby declared to be illegal shall be deemed guilty of a felony, and, on conviction thereof, shall be punished by fine not exceeding $10,000,000 if a corporation, or, if any other person, $350,000, or by imprisonment not exceeding 3 years, or by both said punishments, in the discretion of the court. Section 2 of the Act states that every person who monopolizes, or attempts to monopolize, or combine or conspire with any other person or persons, to monopolize any part of the trade or commerce among the several States, or with foreign nations, would be deemed guilty of a felony, and, on conviction thereof, will be punished by fine not exceeding $10,000,000 if a corporation, or, if any other person, $350,000, or by imprisonment not exceeding 3 years, or by both said punishments, in the discretion of the court.

The watershed event of American competition law could be traced to the *Airline Deregulation Act* of 1978 which liberalized domestic air transport in the US. *The Sherman Antitrust Act* of 1890[18] inculpates conduct that restrain trade and monopolization unlawfully, while *The Clayton Act* of 1914 forbids mergers and

[16] Ioannis Leanos, *Competition law in the European Union after the Treaty of Lisbon*, Academia.edu at https://www.academia.edu/1294134/Competition_Law_in_the_European_Union_After_the_Treaty_of_Lisbon. Article 3.1 (g) provides that For the purposes set out in the preceding Article, the activities of the Community shall include, under the conditions and with the timing provided for in the Treaty the application of procedures which shall make it possible to co-ordinate the economic policies of Member States and to remedy disequilibria in their balances of payments.

[17] *United Brands* v. *Commission*, Case 27/76 [1978] ECR 207 [1978]1 CMLR 429.

[18] 15 U.S.C. SS 1-7, 2.7, 1890.

acquisitions which are calculated to be anti competitive. The teleological basis for both these legal instruments lie in the fundamental postulate of consumer protection which has already been discussed. This principle was enunciated in the 1993 case of *Spectrum Sports Inc* v. *McQuillan*[19] where the court held that the most important aspect of anti competitive trade laws in the country was the need to protect the consumer from the failure of the market.[20] Such practices as discrimination regarding rebates, pricing, services and facilities as well as discounts offered as incentives and "ganging up" with other competitors to eliminate yet others from the market are all prohibited activities under *The Clayton Act.*

10.2.4 Some Conclusions

It seems apparent from the examples cited above that many countries around the world have enacted anti competitive laws, particularly those that preclude State aid and the abuse of dominant position. Yet, when it comes to disadvantages that their national carriers face, they cover themselves with the shield of protectionism. There could be many reasons for that, as seen in the claim of the three American carriers against their open skies partners of the Gulf. Unemployment increases that nationals of the United States may face by the loss of market share caused by Emirates, Etihad and QATAR Airways is one reason given by the American carriers in the docket submitted to the US Congress. The operative question the lifting of trade barriers seems to pose is whether the easing of trade barriers would harm or benefit the local economy. One of the serious challenges posed by this situation is that it cannot be resolved without information that can be disseminated to the interested parties. One commentator opines that since the 1980s, cost benefit information with regard to the lifting of trade barriers and prohibition on subsidies, drawn up by impartial entities and Organizations, reflect that not only developing countries but many developed countries such as Australia and New Zealand have benefitted from open trade. The United States claims, which are quite diametrically opposed to its airlines claims against the Gulf carriers, that "Direct air connections bring substantial economic benefits. Open Skies agreements expand cooperative marketing arrangements, liberalize charter regulations, improve flexibility for airline operations, and include provisions committing both governments to observe high standards of safety and security. These agreements also produce countless new cultural links worldwide[21]".

At the 6th Air Transport Conference of ICAO held in 2012, The United States advised the Conference that liberalization provides broad and deep economic benefits for States' consumers, airlines, airports, communities and economies.

[19] 506 U.S. 447 (1993).
[20] *Id.* 458.
[21] U.S. Department of State, Diplomacy in Action, at http://www.state.gov/r/pa/pl/159347.htm.

Increased access to the international market for air service providers is a key component for allowing the air transport sector to maximize its contribution to the global economy. Furthermore the United States said that air services agreements that couple increased market access (in addition to 3rd and 4th freedom) with the full set of other open-skies elements, including provisions that eliminate government interference in airlines' commercial decisions on capacity and pricing and establish pro-competitive elements on user charges, doing-business matters and cooperative arrangements, deliver broad economic benefits to States and aviation stakeholders. The US additionally advised the Conference that the considerable progress that Member States have achieved in liberalization is evidence that the current system has been very effective in increasing liberalization, and that the momentum should be maintained through the use of open-skies agreements at the bilateral, regional, plurilateral and multilateral levels.[22]

Winston and Yan in a Brookings Institute study published in May 2015 claim that the open skies agreements so far signed by the US have generated at least $4 billion in annual gains to travelers and that travelers would gain an additional $4 billion if the US negotiated agreements with other countries that have a significant amount of international passenger traffic.[23] In the Context of the US-EU Open Skies Agreement, a Brattle Group study conducted in 2002 estimated the value of the agreement at approximately $12 billion annually in economic benefit to the transatlantic airline and related industries. The Study also stated that the agreement would generate 80,000 new jobs.[24] Giovanni Bisignani, the then Director General and CEO of IATA said of the US-EU Open Skies Agreement: "Airlines need greater commercial freedom to run their businesses as real businesses and serve passengers effectively. Today's agreement between the US and the EU is a step in the right direction. I congratulate all involved for moving the agenda forward and challenge them to go even further.[25]"

As the global air transport market is becoming more and more open and competitive, it is more important to ensure that competition is not distorted by unfair practices. And where fair competition conditions exist, it is best to repeal or reduce market access restrictions e.g. in bilateral air service agreements so that airlines can compete freely. Indeed, fair competition is an important principle to achieve the objective of full liberalisation of market access and to reap its benefits. Furthermore, the existence of fair competition is also likely to encourage States to make further progress on liberalising airline ownership and control.

At the Sixth Air Transport Conference of ICAO (*ATConf/6*) States recognized that the principle of fair and equal opportunity is enshrined in the Chicago

[22] Liberalization of Market Access (Presented by the United States of America), ATConf/6WP/60, 14/2/13.

[23] Winston and Yan (2015).

[24] See by Alford and Champley (2007), p. 3.

[25] *IATA Welcomes US-EU Open Skies Agreement*, 22 March 2007, http://www.iata.org/pressroom/pr/pages/2007-03-22-01.aspx.

Convention where States have agreed that international air transport services "may be established on the basis of equality of opportunity" and every State "has a fair opportunity to operate international airlines". What the Conference did not do was to interpret the words "a fair opportunity to operate international airlines". A fair opportunity is not a guaranteed right to operate air services to the detriment of fair competition.

A broad range of issues were raised under the topic of fair competition. A number of States focused on the challenges faced by smaller airlines, especially those from developing countries, when competing against much larger carriers, a challenge made more difficult in some cases by airline mergers and alliances as well as by unilateral or discriminatory measures that deny equitable opportunities. Others focused on issues linked to the inconsistent application of competition laws and policies, including standards for granting antitrust immunity. Some States noted the negative effects on competition caused by barriers to market access. Other States stressed that market liberalization must go hand-in-hand with concrete measures to ensure fair competition.

A number of States supported work by ICAO to establish core principles on fair competition, both to provide a clearer understanding of what is fair and unfair and to indicate appropriate measures to address problems. Among the measures identified were the establishment and effective enforcement of competition laws applicable to international air transport, clear and strong rules on state aids, and the inclusion of appropriate fair competition clauses in air services agreements based on ICAO templates. Other States mentioned principles of fair competition aimed at blocking control of markets by dominant carriers, ensuring all carriers equitable access, prohibiting discrimination and barring abusive practices. However, a number of other States disagreed with the proposal to establish core principles citing (a) that attempts to reach consensus on core principles would prove impossible given the major differences in State views and practices; (b) that issues of fairness can be effectively handled in existing bilateral channels and through use of ICAO template language; (c) that many airlines and airports are State-owned (hence core principles that challenge this fact would be contrary to the principle of State sovereignty in the Chicago Convention); and (d) that core principles on fair competition might be misinterpreted or misused as a barrier to competition.

The Conclusions of the Conference were that in accordance with the Chicago Convention, fair competition is an important general principle in the operation of international air services, and that ICAO policies on competition are still valid, based on observed practices, such as the inclusion of ICAO model clauses on competition in air services agreements. The Conference suggested that ICAO should continue to monitor developments and update its policies and guidance in response to changes in the industry and State practice, and that there is a recognized need for States to give due consideration to the concerns of other States in the application of national or regional competition laws and policies to international air transport as well as a need for cooperation among competition authorities, including in the context of approval of alliances and mergers. In this regard, the Conference was of the view that ICAO should play a leadership role in identifying and developing tools to promote dialogue and the exchange of information among interested authorities with the goal of fostering more compatible regulatory approaches. Such tools could include the development by ICAO of a detailed compendium of national and regional competition policies and practices as well as the development of a facility that would serve as a forum for the enhancement of cooperation, dialogue and exchange of information.

Accordingly *ATConf/6* recommended as follows:

a) States should take into consideration that fair competition is an important general principle in the operation of international air services;
b) States, taking into account national sovereignty, should develop competition laws and policies that apply to air transport. In doing so, States should consider ICAO guidance on competition;
c) States should give due consideration to the concerns of other States in the application of national and/or regional competition laws or policies to international air transport;
d) States should give due regard to ICAO guidance in Air Services Agreements (ASAs) and national or regional competition rules;
e) States should encourage cooperation among national and/or regional competition authorities, including in the context of approval of alliances and mergers;
f) ICAO should develop tools such as an exchange forum to enhance cooperation, dialogue and exchange of information between Member States to promote more compatible regulatory approaches toward international air transport;
g) ICAO should develop a compendium of competition policies and practices in force nationally or regionally; and
h) ICAO should continue to monitor developments in the area of competition in international air transport and update, as necessary, its policies and guidance on fair competition through the Air Transport Regulation Panel (ATRP).

The only recommendation that stands out as being different on a general basis from those in *ATConf/5* is the suggestion that ICAO develop a compendium of competition policies and practices in force nationally or regionally. In developing these policies, ICAO could do well to take into consideration the fact that fair competition is a key principle to achieve the benefits of liberalisation of market access in international air transport at worldwide level and that States should take measures to ensure fair competition, for example, through laying down efficient competition laws applicable to international air transport, as well as clear, transparent and strict state aid rules, developing and inserting fair competition clauses in their bilateral air services agreements, and through closer cooperation between their respective authorities including, where appropriate, in competition investigations. Critical to ICAO's considerations would be the possibility of introducing a fair competition clause to be included in bilateral air services agreements to establish and maintain it at global level.

What is needed in the dispute between the US carriers and Gulf carriers is a number crunching exercise on how the surge by the consumer of the air transport product toward services offered by the Gulf carriers in the US market is benefitting the overall industry, the air transport market and ultimately the consumer. One cannot ask the Gulf carriers to change their business models and their advantageous geographic locations. But one can certainly ask them to desist from any anti-competitive conduct. So far, no one has claimed that these carriers are guilty of anti competitive conduct and are acting illegally. The Gulf carriers are merely exercising their rights under prevailing open skies agreements. It is just that one side is winning, and the other side seems to be losing. That, in essence, is competition in air transport where protectionism still seems to be rampant.

References

Abeyratne R (2012a) Administering the skies – facing the challenges of market economics. Aracne Editrice, Rome

Abeyratne R (2012b) Aeronomics and law: fixing anomalies. Springer, New York

Abeyratne R (2013a) Aviation and climate change: in search of a global market based measure. Springer, Heidelberg

Abeyratne R (2013b) Regulation of air transport: the slumbering sentinels. Springer, New York

Alford E, Champley R (2007) The impact of the 2007 U.S.-EU open skies air transport agreement. ITA Occasional Paper no. 07-001, May, p 3

Anderson K (2004) Subsidies and trade barriers. In: Global crises. Global Solutions, Cambridge, p 544

Porter ME (1998) On competition. Harvard Business Review Book Series, p 184

Truxal S (2012) Competition and regulation in the airline industry: puppets in chaos. Routledge, pp 93–102

Winston C, Yan J (2015) Open skies: estimating travelers' benefits from free trade in airline services, http://www.brookings.edu/research/articles/2015/05/13-open-skies-agreement-passenger-gains-winston

Chapter 11
Air Transport and the Law of Investment

11.1 Introduction

Investment in airlines, airline fleets and other resources is a key driver of air transport. The Airline Pilots Association (ALPA) issued a statement accusing the government of the United Arab Emirates of investing in Dubai International Airport, implying that it was a subsidy granted to Emirates Airlines and that such an act flouted the "fair and equal opportunity to compete" clause in the open skies agreement between the United States and the UAE.[1] The Fact Sheet in the public domain provided by the Department of State says about open skies:

> Open Skies agreements between the United States and other countries expand international passenger and cargo flights by eliminating government interference in commercial airline decisions about routes, capacity and pricing. This frees carriers to provide more affordable, convenient and efficient air service to consumers, promoting increased travel and trade and spurring high-quality job opportunity and economic growth. Open Skies policy rejects the outmoded practice of highly restrictive air services agreements protecting flag carriers.[2]

There is nothing in this statement about fair and equal opportunity to compete in operating air services. The Fact sheet goes further to say that before Open Skies began to liberalize the international aviation environment, cities like Dallas-Fort Worth, Detroit, Las Vegas, Memphis, Minneapolis, Portland, and Salt Lake City had few or no direct international air connections. Now they enjoy direct connections to cities around the world.[3]

It must be noted that the Model Open Skies Agreement of the US states that Each Party has to allow a fair and equal opportunity for the airlines of both Parties to

---

[1] *ALPA*: *Gulf Airline Subsidies Have No Parallel in U.S.*, http://www.atn.aero/article.pl?id=55544.

[2] Open Skies Partnerships: Expanding the Benefits of Freer Commercial Aviation, Fact Sheet, Bureau of Public Affairs, March 29, 2011. See http://www.state.gov/r/pa/pl/159347.htm.

[3] *Ibid.*

R. Abeyratne, *Competition and Investment in Air Transport*,
DOI 10.1007/978-3-319-24372-6_11

compete in providing the international air transportation governed by this Agreement. However Clause 2 of the Model Agreement goes on to say that each party is required to allow each airline to determine the frequency and capacity of the international air transportation it offers based upon commercial considerations in the marketplace. Consistent with this right, neither Party has the right to unilaterally limit the volume of traffic, frequency, or regularity of service, or the aircraft type or types operated by the airlines of the other Party, except as may be required for customs, technical, operational, or environmental reasons under uniform conditions consistent with Article 15 of the Chicago Convention.[4] Also, the following clause states that neither party has the right to impose on the other party's airlines a first-refusal requirement, uplift ratio, no-objection fee, or any other requirement with respect to capacity, frequency, or traffic that would be inconsistent with the purposes of the Agreement.

The issue with the ALPA claim against the investments in Dubai International Airport preventing US carriers the right to fair and equal opportunity to compete in operating air services is somewhat puzzling. How does improvement of airport infrastructure and facilities in one country prevent carriers from another country from competing in operating air services. The ALPA claim has received an implicit nod in an official US Congressional Research Service (prepared for the US Congress) document issued in May 2015 which recognizes with some concern:

> [T]he government of Dubai is building a new airport, Al Maktoum International Airport, which it envisions as the largest passenger hub in the world. Al Maktoum, located approximately 40 miles from Dubai International Airport, is used mainly as a cargo airport at present and sees only a handful of passenger flights. Last September, Dubai's ruler, Sheik Mohammed bin Rashid Al Maktoum, approved a $32 billion expansion plan that would enable the airport to handle 120 million passengers per year and to service 100 Airbus 380 jets at the same time. Further expansion plans would take capacity to 220 million passengers per year. The government expects Emirates to relocate its hub to the new airport. The existing Dubai International Airport is expected to remain in operation.[5]

---

[4] Article 15 states that every airport in a contracting State which is open to public use by its national aircraft shall likewise. . .be open under uniform conditions to the aircraft of all the other contracting States. The like uniform conditions shall apply to the use, by aircraft of every contracting State, of all air navigation facilities, including radio and meteorological services, which may be provided for public use for the safety and expedition of air navigation. Any charges that may be imposed or permitted to be imposed by a contracting State for the use of such airports and air navigation facilities by the aircraft of any other contracting State shall not be higher, (a) As to aircraft not engaged in scheduled international air services, than those that would be paid by its national aircraft of the same class engaged in similar operations, and (b) As to aircraft engaged in scheduled international air services, than those that would be paid by its national aircraft engaged in similar international air services. All such charges shall be published and communicated to the International Civil Aviation Organization: provided that, upon representation by an interested contracting State, the charges imposed for the use of airports and other facilities shall be subject to review by the Council, which shall report and make recommendations thereon for the consideration of the State or States concerned. No fees, dues or other charges shall be imposed by any contracting State in respect solely of the right of transit over or entry into or exit from its territory of any aircraft of a contracting State or persons or property thereon.

[5] Tang (2015), p. 9.

There is no document, agreement or treaty that ties in government investment in investment in airport capacity and infrastructure to a foreign carrier's inability to compete fairly and equally in air transport. How does development of State infrastructure distort a level playing field? Would this not be like saying that two candidates at an job interview should be given the same background information to face the interview and one should be prevented from investing in books to enhance his chances of being selected for the job?

The Chicago Convention in its Preamble speaks of air transport services being operated with *equality of opportunity*, without elaborating further. A theme often sounded in the discussion of international air transport competition is the need for a "level playing field". On an abstract level, there is broad acceptance of the principle that a fair and equal opportunity is required to allow airlines to succeed and grow in the liberalized global market. In fact, some air services agreements refer explicitly to the principle of a level playing field by noting that "where there is not a level competitive playing field for airlines, potential benefits deriving from competitive air services may not be realised". However, it must be recognized that there is currently no commonly accepted definition of the conditions constituting a "level playing field". It is unlikely that consensus on a comprehensive definition can be achieved at this time, given the widely different circumstances of States and their aviation sectors, including such fundamental issues as state ownership, policies on maintenance of national air carriers and airport development, and widely divergent State policies on taxation, labour regulation, bankruptcy, and health insurance.

11.2 Competition and Investment in Air Transport Infrastructure

At the time of writing, it had been reported that in 2014, countries in the Asia Pacific region were planning to spend billions of dollars into building airport infrastructure to handle a higher-than-expected demand for air transport in the region bringing a substantial influx of passengers to the region The same report said that China, India, Philippines and Indonesia, were planning to invest over $100 billion in new airports over the next decade to accommodate burgeoning passenger traffic in the region.[6] One commentator said: "Through the next 10 years, we will see more than 350 new airports in the Asia-Pacific region and the investment cost will be well over US$100 billion[7]". He added "China is building over 100 airports,

[6] Pandey (2014). See http://www.ibtimes.com/asian-countries-spend-100b-new-airports-passenger-traffic-grows-report-1644414.

[7] De Lavigne (2014). According to the United Nations World Tourism Organization, tourism in the Asia Pacific region grew at a yearly pace of over 6 % in the first 4 months of 2014—higher than the global average of 5 %. Within the Asia Pacific region, the growth was much higher in South Asian countries, which witnessed a yearly growth of over 8 %.

India is building over 60 airports and Indonesia will also have to follow suit with investments in its infrastructure".

It is noteworthy that in 2014, China had earmarked $11 billion for a new airport in Beijing, and India had announced plans to build 200 low-cost airports between 2015 and 2035. Indonesia is planning to the upgrade of Soekarno-Hatta International Airport in the capital city of Jakarta, which is currently running at three times its normal capacity. In Singapore, Even Changi Airport in Singapore, which currently handles over 54 million passengers annually, recently announced a series of measures to increase passenger traffic further, including investing $1 billion in building a new terminal. *The Global Infrastructure Index* of 2014 states:

> Ports and airports generally do not give the same guarantees of returns but the understanding of freight and passenger traffic that is inherent with aviation satisfies the analytic approach sought by investors and can make these assets a sound proposition. A number of airports internationally have proven the model can be profitable from BAA in the UK to Macquarie in Sydney and airports in Europe, Asia and the Middle East are all considering more private involvement... Access to finance will be critical as these countries scale up investment. In Qatar and the UAE in particular, national vision strategies combined with major international events have led to expectations of phenomenal peak spend in the next 4–5 years. Almost half of the investment planned relates to transportation, with every major city in the region planning to follow Dubai in building a metro system with lines being constructed simultaneously in a relatively short period of time. At the same time ports, airports and a heavy rail network are all under construction leading to increased competition for resources. The key risk in these markets therefore is inflation in construction resources from manpower and specialist skills to construction commodities.[8]

Oxford Economics makes the point that The Dubai government and Emirates Airlines airline have played a mutually supportive role in leading Dubai's economic development:

> The relationship between the government and the aviation sector is a consensus-based partnership that allows important decisions to be made quickly and carried through effectively. The success of the partnership is dependent on the government's awareness of the importance of aviation to Dubai's economy and the shared vision provided by the Strategic Plan of the Dubai Government.[9]

This view supports the key point made by the *Global Infrastructure Index*—that investment in the development of aviation infrastructure in the Gulf countries is essentially due to political will and governmental acknowledgment of the need for such development for economic growth. In contrast, the Index talks of governmental reluctance and sloth with regard to investment in development in Europe.

> ...Holland, France and Germany have taken a very assertive approach by expanding airports such as Amsterdam's Schipol and Paris' Charles De Gaulle. However, these countries have other issues to contend with. Sitting in the middle of the table, Western European countries, in general, are lacking government commitment to progress projects which remain in the pipeline and need to upgrade their ageing infrastructure. This leaves

[8] *Second Global Infrastructure Index for 2014*, Competing for Private Finance, Arcadis:2014 at 23.

[9] *Explaining Dubai's Aviation Model, A Report for Emirates and Dubai Airports*, Oxford Economics: June 2011, at 5.

> them with the choice of doing nothing, which means further deterioration in condition and effectiveness of infrastructure, or they could stimulate longer term private finance. This could mean either embarking upon a series of part privatisations akin to the UK experience or following the example of cities such as Los Angeles and Chicago who have used their credit worthy status to raise bond issues to spend on infrastructure investment plans.[10]

Against this backdrop, and the fact that infrastructure in the United States badly needs an overhaul, it is tenuous to argue that investing in growth and development in aviation, particularly in the airport sector in Gulf countries pose an uneven playing field for the US carriers, when no attempt has been made in the US in terms of investing in aviation infrastructure. The American Society of Civil Engineers has, in its report on aviation infrastructure in the US has given it a D- saying that despite the effects of the recession in 2008–2009, commercial enplanements were about 33 million higher in number in 2011 than in 2000, stretching the system's ability to meet the needs of the nation's economy. The Federal Aviation Administration (FAA) estimates that the national cost of airport congestion and delays was almost $22 billion in 2012. If current federal funding levels are maintained, the FAA anticipates that the cost of congestion and delays to the economy will rise from $34 billion in 2020 to $63 billion by 2040.[11]

11.3 Attribution

The issue of attribution is central to the question as to a level playing field in competition for market access in air transport can be affected by a State's development of internal infrastructure within that State. Another way of looking at the question is to inquire as to whether competition in air transport in the context of fair and equal opportunity extends from airline practice of anti competitive conduct to the responsibility of a State to facilitate economic growth in general. A Report published in April 2015 by *Airline Information and Thinkink* says that the key factors that will influence airlines in 2015 and beyond are improved passenger experience with a focus on experience, including initiatives and strategies that make flying worthy; continued and re-energized focus on revenue brought to bear by enhanced partnerships, ancillary revenue opportunities and products and services passengers want; streamlined and seamless technologies; enhanced loyalty focusing on benefits and rewards; and continued growth of mobile payments.[12] One does not see airport and other aviation infrastructure investments in this list. On the other hand, the customer seeks a safe personalized way to book travel anywhere from anywhere, to be present wherever the customer is, and these factors are

---

[10] *Id.* 16.

[11] *2013 Report Card for America's Infrastructure*, The American Society of Civil Engineers:2014 at http://www.infrastructurereportcard.org/a/#p/bridges/overview.

[12] *Flying Forward: What the Future Holds for Airlines in the Travel Industry*, Airline Information and Thinkink, April 2015, at 5.

common to most competing airlines whether they are in the United States or in the Gulf or in Australia.

The main issue is whether a State's investments on economic growth can be attributed to the State as an anti competitive practice hinges on the doctrine of attribution and State responsibility. A good case in point to illustrate attribution and State responsibility is Emirates Airlines and the government of Dubai. A starting point is the philosophy of the Dubai Government which the Oxford Economics Study identifies as giving equal benefits to all airlines operating into Dubai—not only to Emirates Airlines. The Report goes on to explain that the fact that Dubai favours and indeed seeks open competition among airlines is reflected by the equal benefits gained by the improvements made to Dubai International Airport that are enjoyed by over 150 airlines operate out of Dubai International, benefiting from investment in aviation infrastructure and competitive landing charges. Moreover, to further facilitate the open skies concept in its true notion of equality and fairness, the Report claims that Dubai's Civil Aviation Authority has pushed for greater freedoms for all airlines to enable them to operate without undue restrictions on their commercial decisions. Dubai eschews restrictions of market access rights elsewhere and has sought, through negotiation with other governments, to improve this situation.[13]

Dubai Airports is directly owned by the government but Emirates Airlines is owned by Investment Corporation of Dubai which is a government owned investment company. This notwithstanding, Emirates is fully corporatized and operates autonomously without any financial support from or control of the government. This commercial situation, between the government and a corporatized airline is not unique to Dubai and can be found as a common model elsewhere. State responsibility of the Dubai Government is to enhance the economy of the country by investing in the infrastructure without unduly favouring its national carrier over other airlines operating into Dubai, and the responsibility of the airline is to concentrate on the key factors that would influence its profitability and success as shown the Report of *Airline Information and Thinkink*. It is in this context that Emirates Airlines can stand alone, where, although it is owned by a company that is wholly government owned, the government of Dubai considers it as an independent business entity.[14] Article 8 of the International Law Commission's Articles on State Responsibility states that the conduct of a person or group of persons shall be considered an act of a State under international law if the person or group of persons is in fact acting on the instructions of, or under the direction or control of, that State in carrying out the conduct. In the case of Emirates Airlines and the Government of Dubai, an interpretation of Article 8 of the ILC principles would show that the two entities are mutually exclusive either way in terms of responsibility and therefore the injection of investment by the Dubai government to airport infrastructure cannot be ascribed to the commercial conduct of the airline.

---

[13] *Ibid.*

[14] http://www.emirates.com/english/about/the_emirates_story.aspx.

The entire principle of linkage between government entities and government lies on the element of control. The International Court of Justice, in the 1986 *Nicaragua case* held that the conduct of the Contras could not be attributed to the United States despite the heavy subsidies and other support provided to them by the United States, as there was no clear evidence of the United States having actually exercised such a degree of control in all fields as to justify treating the Contras as acting on its behalf.[15] The link is further strengthened by eminent commentators who state:

> Therefore, attribution to the state of conduct under the "direction or control" of the State requires not only that the entity is generally controlled by the state but that the individual operation in question was effectively controlled and that the act was a genuine part of that operation.[16]

Under principles of international law, therefore, it would be tenuous to draw a link between government investment in Dubai International Airport and the commercial conduct of Emirates Airlines. Judicial determination of activities of State entities or alleged State entities as is the case of Emirates Airlines has essentially been decided on the criterion of functionality. In the 2005 *Maffezini* case[17] the court pronounced that reliance must be placed on the functionality test and establish whether specific acts in question are essentially commercial in nature or on the other hand they were governmental in nature, or whether the nature of the acts in question are essentially governmental rather than commercial. Although in the same year, a judgment[18] differed with the *Maffezini* ruling, on the ground that there was no reason why *acta jure gestionis* or commercial acts are not attributable, and that it was difficult to define what a government act was, considered opinion is that the ILC Article clearly states that attribution, regardless of the nature of the act applies only to State organs.

Another economic reality about competition in the context of national competitiveness is that National competitiveness is one of the most critical drivers of successful government and industry in every nation. Yet for all the discussion, debate, and writing on the topic, there is still no persuasive theory to explain national competitiveness. What is more, there is not even an accepted definition of the term "competitiveness" as applied to a nation. While the notion of a competitive company is clear, the notion of a competitive nation is not. The deliberations of the Sixth Air Transport Conference of ICAO clearly brought to bear this point and implicitly called upon States to revisit their own strategies with regard to the air transport industry on the basis that the most important feature of a competitive nation is its decisive characteristic that allows its companies to create and sustain competitive advantage in particular fields—the search is for the competitive advantage of nations. Of particular concern are the determinants of

---

[15] *Military and Paramilitary Activities in and against Nicaragua* (*Nicaragua* v. *United States of America*), *Merits*, *Judgment*, *I.C.J. Reports 1986* p. 51, para. 86.

[16] Dolzer and Schreuer (2008), p. 201.

[17] *Emilio Augustin Maffezini* v. *The Kingdom of Spain*, ICSID Case No. ARB/97/7, para. 52.

[18] *Noble Ventures* v. *Romania*, Award, 12 October 2005.

international success in technology and skill-intensive segments and industries, which underpin high and rising productivity.

Classical theory supports the principle that the success of nations in particular industries based on so-called factors of production such as land, labor, and natural resources is based on the fact that nations gain factor-based comparative advantage in industries that make intensive use of the factors they possess in abundance. Classical theory, however, has been overshadowed in advanced industries and economies by the globalization of competition and the power of technology.

Any new approach must recognize that in modern international competition, companies compete with global strategies involving not only trade but also foreign investment. What a new theory must explain is why a nation provides a favorable home base for companies that compete internationally. The home base is the nation in which the essential competitive advantages of the enterprise are created and sustained. It is where a company's strategy is set, where the core product and process technology is created and maintained, and where the most productive jobs and most advanced skills are located. The presence of the home base in a nation has the greatest positive influence on other linked domestic industries and leads to other benefits in the nation's economy. While the ownership of the company is often concentrated at the home base, the nationality of shareholders is secondary.

A new theory for air transport must seemingly move beyond comparative advantage to the competitive advantage of a nation. It must reflect a rich conception of competition that includes segmented markets, differentiated products, technology differences, and economies of scale. A new theory must go beyond cost and explain why companies from some nations are better than others at creating advantages based on quality, features, and new product innovation. A new theory must begin from the premise that competition is dynamic and evolving; it must answer the questions: Why do some companies based in some nations innovate more than others? The most important feature of a competitive nation is its decisive characteristic that allows its companies to create and sustain competitive advantage in particular fields—the search is for the competitive advantage of nations. Of particular concern are the determinants of international success in technology and skill-intensive segments and industries, which underpin high and rising productivity.

References

De Lavigne C (2014) Hundreds of new airports to meet Asia tourism boom. India Post, August 5, http://www.indiapost.com/hundreds-of-new-airports-to-meet-asia-tourism-boom/

Dolzer R, Schreuer C (2008) Principles of international investment law. Oxford, New York, p 201

Pandey A (2014) Asian countries to spend $100B on new airports as passenger traffic grows: report. International Business Times, July 31

Tang RY (2015) International air service controversies: frequently asked questions. Congressional Research Service 7-5700 www.crs.gov., R44016, May 4, p 9

Chapter 12
Conclusion

The trouble with competition in air transport is that the main guiding light—ICAO—functions like a ship without a rudder. The most recent example of this phenomenon occurred in mid-2015 at the time this book was being written. To perpetuate ICAO's ambivalence in its "leadership" role in air transport, the ICAO Council, in July 2015, misdirected the Member States of ICAO, by pursuing them to adopt the following grandiose long-term vision for the liberalization of air transport:

We, the Member States of the International Civil Aviation Organization, resolve to actively pursue the continuous liberalization of international air transport to the benefit of all stakeholders and the economy at large. We will be guided by the need to ensure respect for the highest levels of safety and security and the principle of fair and equal opportunity for all States and their stakeholders.

The question that immediately arises is "fair and equal opportunity" in what respect? And to do what? The Chicago Convention clearly and explicitly states that it is air services that should be operated fairly with equality of opportunity. The word "opportunity" is key and the wisdom of the founding fathers of this treaty should not go un-acclaimed. Opportunity is a set of circumstances that makes it possible to do something. In other words, it is clear that the Chicago Convention speaks of fair and equal opportunity of an airline to compete. Would it have been too difficult for ICAO to have at least considered this simple fact and mentioned competition in the equation?

Should not ICAO have heeded The Sixth Worldwide Air Transport Conference (ATConf/6, 2013)'s Recommendation 2.4/1, which addressed, among other key issues, the topic of fair and equal opportunity adopted a conclusion which calls for States to consider fair competition as an important general principle in the operation of international air services? In the field of air transport, clarity of language has never been ICAO's virtue or strong point. However, this does not mean that some attempt should not be made to be unequivocal and articulate and purpose of a statement even though it results in the wrong conclusion being reached. This vague vision was adopted in pursuance of the Council's consideration of the fact presented

R. Abeyratne, *Competition and Investment in Air Transport*,
DOI 10.1007/978-3-319-24372-6_12

to it by a Secretariat working paper which said that passengers can benefit from a competitive air transport sector, which offers more choice in fare-service trade-offs and which may encourage carriers to improve their offerings, passengers, including those with disabilities, can also benefit from consumer protection regimes. Therefore it was ICAO's contention that Government authorities should have the flexibility to develop consumer protection regimes[1] which strike an appropriate balance between protection of consumers and industry competitiveness and which take into account States' different social, political, and economic characteristics, without prejudice to the safety and security of aviation. National and regional consumer protection regimes should: reflect the principle of proportionality; and allow for the consideration of the impact of massive disruptions.

Such policies, it was claimed, should be consistent with the international treaty regimes on air carrier liability established by the Convention for the Unification of Certain Rules Relating to International Carriage by Air (Warsaw, 1929) and its amending instruments, and the Convention for the Unification of Certain Rules Relating to International Carriage by Air (Montréal, 1999).

One of the problems of interpretation ICAO faces is the maundering of the member States themselves. The Air Transport Regulation Panel, which held its meeting in 2014 recommended "aspirational and not prescriptive in nature, should emphasize a commitment to liberalization, and should be forward looking, representing a long term view of the aviation community on the ultimate objective of liberalization. The questions would be: If such policy were aspirational and non-prescriptive, what coercive force would it have in directing States toward competition and liberalization? Also what is the purpose of a "long term" view of liberalization"? How long would States have to wait before liberalizing air transport?

Here the ICAO Council has made the usual blunder of equiperating principles of private air carrier liability, as enshrined in the multilateral treaties mentioned above, with consumer protection. Whereas the said treaties offer compensation after the fact, i.e. for death or injury caused by an accident, or delay which occurs in the course of transportation by air, consumer protection is involved with protecting the consumer before an irregularity or injury occurs. This unfortunate confusion having been deposited in the minds of the Council by the ICAO Secretariat, it reverts back to consumer protection in the Working Paper by saying that it is recognized that "the variety of air transport products in the market, passengers should have access to information on their rights and clear guidance on which legal or other protections apply in their specific situation, including the assistance expected, for example, in case of service disruption. To help air passengers make informed choices among different price and service offerings, consumer education efforts could be considered to increase awareness of passengers consumer rights and the available avenues for recourse should disputes arise. Efforts should also be

[1] *C-WP/14269, Appendix A.*

made to increase awareness by passengers of airline products available in the market, different airline policies and contractual rights".

The Council is further briefed (in the Working Paper) that passengers should have clear, transparent access to all pertinent information regarding the characteristics of the air transport product that is being sought, prior to purchasing the ticket, including the total price, including the applicable air fare, taxes, charges, surcharges and fees; general conditions applying to the fare; and identity of the airline actually operating the flight, and advice on any change occurring after the purchase as soon as possible.

Another glaring circumlocution in the Council's consideration was the recommendation that passengers should be kept regularly informed throughout their journey on any special circumstances affecting their flight, particularly in the event of a service disruption. What are these special circumstances? Should the passengers be informed of details that are usually issued in Notices to Airmen (NOTAMs)[2]? For example, should the passenger necessarily know during flight that the pilot has skilfully avoided flying over a conflict zone? Of what use would this information be to the passenger? Would the passenger be better off if she did not have this information but instead was given information that would practically ensure her safety inflight, such as how to respond to turbulence?

Arguably, the weakest link in the Council's deliberations which were guided by the feckless insouciance of directions of the ICAO Secretariat in the Working paper, is the confounded association of consumer protection and competition. Whether or not there is liberalization or competition, consumer protection is a right that has to be enforced. On the other hand liberalization is a commercial tool that ensures economic efficiency. These two should have been separately presented and discussed.

There are also sensibilities to be addressed. At the World Wide Air transport Conference convened by ICAO in March 2013 the United Arab Emirates urged ICAO to advance and strengthen its vital work on liberalization in air transport for the benefit of the world; and requested ICAO to continue assisting States by providing studies that analyze the effects of liberalization.[3] One is yet to see these studies 2 years on. The most compelling need remains ICAO's mindset to change after more than 70 years of its existence which would link liberalization with competition in air transport. If ICAO were to attain its aim under the Chicago Convention—of fostering the development of air transport—it has to transcend the simplistic label of "liberalization"—that states should liberalize as they please—to guiding States as to how to compete in accordance with the principles of the Chicago Convention. The furthest ICAO has gone is to use the concept of competition in the Preamble of its Core Principles in Consumer Protection (and not in

[2] "International notices to airmen" is a phrase which led to the birth of an early aeronautical acronym: NOTAM. Specifications of a NOTAM are prescribed in Annex 15 to the Chicago Convention.

[3] Advancing ICAO's Work on Air Transport Liberalization, *ATConf/6-WP/33*, 12/2/13.

liberalization of air transport) saying that ICAO member states recognize that passengers can benefit from a competitive air transport sector, which offers more choice in fare-service trade-offs and which may encourage carriers to improve their offerings, passengers, including those with disabilities, can also benefit from consumer protection regimes. One is baffled by the fact that the term "competition" does not appear in ICAO's vocabulary associated with liberalization.

Irrespective of the regulatory deficiencies air transport faces, airlines have to be aware of the factors which affect competition. The airline industry is heavily oligopolistic, where three large airline alliances i.e. SkyTeam (made up of Delta, which had merged with Northwest; the merged Air France and KLM and Alitalia), Oneworld (British Airways and American) and Star Alliance (United, which merged with Continental; Lufthansa and Air Canada). Together, these three major alliances control or provide 85 % of transatlantic traffic and fares and 50 % of global fares. Allan Mendelsohn, a former Deputy Assistant Secretary of State of the US State Department had this to say:

> ...It does not seem that the current system is all that different from IATA rate making machinery of years ago, when a select group of international airlines participated and set rates within IATA's then rate-making conferences.

Strategic alliances of the airline industry is but a natural corollary to the exponential growth of international air transport as an industry. The concept itself is based on the theory that with rapid demand for air transport, requiring a doubling of the 16,000 world aircraft fleet by the year 2015, these would be a compelling need for new connections between points and more frequencies to serve these connections. There is no stopping this trend, which has already swept the aviation industry. There is, however, one point of caution. The fundamental postulate of air transport has been, and remains to be, safety of passengers. The proliferation of aircraft in the skies may challenge airline safety, if parallel measures are not set in motion to ensure the safe passage of the thousands of aircraft in the sky.

As we are headed towards the next decades, this situation brings to bear the compelling need for corporate social responsibility that would ensure fair and equal opportunity for airlines to compete in providing air transport without being controlled by external factors dictated to by oligopolies. The necessity for corporate foresight stems from the continuing and rapid development of science and technology which are the drivers of social and economic change. Using these two knowledge based and fact intensive fields, airports would be able to obtain a clear picture of challenges and opportunities confronting them. Airlines are a complex, big business and their business environment is highly dynamic. Therefore they need proactive measures to respond to the uncertainties of their business as well as a long term orientation to remain stable amidst imponderables. Airlines need think tanks to mesh their technology trends and market trends to meet a growing demand for air travel. Foremost in this process is a far reaching and forward looking communications strategy as well as a good team of scientific and economic forecasters.

Competition in air transport lies in entrepreneurship and innovation. Business magnate Rupert Murdoch has said that intuition, innovation and opportunism are hallmarks of competitive advantage and success.[4] This has to be buttressed by State support, as in the example of Silicon Valley, companies of which owe their stature and competitive success to Government funding of Stanford, Berkeley and Caltech.[5] It must be noted that for intuition and opportunism, there has to be corporate foresight.

The first step to corporate foresight is to know what the future is going to be like by adopting a foresight-awareness culture. If, as Airport Council International (ACI) Director Angela Gittens said at the 20th World Annual General Meeting of ACI in Bermuda in October 2010, airports should transition from the public utility model to the entrepreneurial business model, the key would be customer service excellence. Research and innovation strategies should necessarily be developed through foresight activities. This analogy applies equally to airlines. "Foresight" has been defined as:

[a] participatory, future intelligence gathering and medium-to-long-term vision-building process that systematically attempts to look into the future of science, the economy and society in order to support present-day decision-making and to mobilise joint forces to realise them.

Corporate foresight is a process of formulation and should not be confused with a set of techniques. Through a sustained consultative process, corporate foresight involves the examination of a series s of future scenarios and ideally prescribes solutions. Foresight shares common ground with risk management and evaluation of risk, and addresses the nature of the particular business and the uncertainties of the business environment. For example, in the airport industry, a grave uncertainty is the weather, as in the context of an unexpected winter storm or the eruption of a volcano which spews ash into the atmosphere. Both these events occurred in 2010 in Europe.

Tony Tyler, Director General of IATA said at the 4th Passenger Symposium in October 2014 that aviation is built on collaboration and that every flight requires a choreography of cooperation involving airlines, airports, distribution and travel technology providers, travel sellers and air navigation service providers. In 2012 Tyler made a link between the potential of air transport and its contribution, which goes to show the importance of competition in meeting the exponentially growing demand of air transport. Tony Tyler had this to say at the Annual General Meeting of IATA held in Beijing in June 2012, As reported in Air Transport News:

> The major benefits of aviation such as connectivity and economic welfares are evident in tremendous contributions of aviation to the global economy such as the provision of 57 million jobs worldwide. However, the state of the industry has been characterized as fragile, with 631 billion dollars revenue but only 3 billion dollar profit for 2012 with high oil prices, political risks around the world and the crisis in the Eurozone a large number of airlines is struggling to keep revenues ahead of costs. Governments and airlines should

[4] http://www.amadeus.com/web/amadeus/en_SY-SY/Amadeus-Home/News-and-events/Events/2014-10-15-World-Passenger-Symposium-2014/1319550590434-Page-AMAD_DetailPpal?assetid=1319606989208&assettype=Eventable_C.

[5] *Id.* 36.

> work as strong partners in order for modern economies to grow prosper and create jobs through being exposed to global opportunities.[6]

Ironically, one could place these facts against the backdrop of pronouncements of high level policy makers who are increasingly reaching the conclusion that aviation is making a substantial contribution to the global economy. Oxford Economics, an economic forecasting agency, in a recent report recognizes the wide range of benefits that air transport brings to economies and societies globally.

This report, issued in June 2009, also suggests that the world's future prosperity may depend on a growing and thriving aviation industry, which currently supports nearly eight per cent of the world's economy, and questions the environmental benefits and social impacts of limiting that growth. One wonders, then, as to why an indispensable economic tool such as the air transport industry which contributes so substantially to the global economy, is shackled by restrictions which other similar businesses are not subject to. It is hoped that the discussion below would shed some light on the issue.

IATA has made the point that air transport has to be treated like any other business. If this principle is recognized as a reality, there would not be such a desperate need for airlines to seek alliances in the proportions they do now.

The main consideration, leading up to efforts by the international aviation community to achieve a deregulated global airline industry, is involved with the question as to whether free market principles can be applied globally to air transport. What needs to be considered is whether we are ready to accept the throwbacks as consequences of free market competition in air transport, particularly in losing national prestige projected by flag carriers. One of the corollaries to industry deregulation is the introduction of free market competition when companies switch from operative performance to competitive performance. Competition therefore emphasizes the need to focus on a company's performance in relation to its competitors. This principle can be readily apply to various industries that have already been deregulated, such as the motor vehicle industry, chemical industry and information technology industry. The operative question is "are these good analogies for application to the air transport industry?" Whatever be the answer to this question, if the deregulated domestic air transport industry of the United States were to be considered an analogy, one could say that a deregulated system in the United States, introduced in 1978, has led to a more efficient airline system in the country. Whatever be the case, access to facilities in a competitive market is essential toward attaining fluidity of market forces. In the air transport industry, this can be translated to mean that if free markets do not exist in the supply of complementary facilities, there will be no positive impact of liberalization. The complementary services in the supply of air transport are airport access, computer reservation systems and airport and air regulation services.

[6] http://www.amadeus.com/web/amadeus/en_SY-SY/Amadeus-Home/News-and-events/Events/2014-10-15-World-Passenger-Symposium-2014/1319550590434-Page-AMAD_DetailPpal?assetid=1319606989208&assettype=Eventable_C.

The International Chamber of Commerce (ICC), in a policy statement has expressed the view that the efficiency of air transport would be enhanced by creating more open markets and more flexibility with regard to foreign ownership. Given air transport's capability to facilitate economic activity, its liberalization would enable the sectors that make use of it to become non efficient. ICC was in favor of a freer exchange of air services throughout the world and is convinced that it is time to move beyond the existing bilateral system, toward a genuine multilateral liberalization of air transport. Of course, liberalization would give way to competition, which in turn would impel airlines to pool their resources in order that they maximize on such assets as code sharing and airport slots. However, alliances do not necessarily mean lack of competition between partners. Airlines within alliances have to do their utmost to gain market access and keep their businesses alive. In order to do this both private enterprises and the States in which these enterprises are entrenched have to be equally competitive.

Any agreement to bring in an aspect of trade within a liberalized framework is generally a pro-active measure, which brings to bear the willingness and ability of the governments to face trading issues squarely in the eye. However, any agreement for trading benefits would be ineffective without the element of translated to mean that if free markets do not exist in the supply of complementary facilities, there will be no positive impact of liberalization. The complementary services in the supply of air transport are airport access, computer reservation systems and airport and air regulation services.

The International Chamber of Commerce (ICC), in a policy statement has expressed the view that the efficiency of air transport would be enhanced by creating more open markets and more flexibility with regard to foreign ownership. Given air transport's capability to facilitate economic activity, its liberalization would enable the sectors that make use of it to become non efficient. ICC was in favor of a freer exchange of air services throughout the world and is convinced that it is time to move beyond the existing bilateral system, toward a genuine multilateral liberalization of air transport Of course, liberalization would give way to competition, which in turn would impel airlines to pool their resources in order that they maximize on such assets as code sharing and airport slots. However, alliances do not necessarily mean lack of competition between partners. Airlines within alliances have to do their utmost to gain market access and keep their businesses alive. In order to do this both private enterprises and the States in which these enterprises are entrenched have to be equally competitive.

Any agreement to bring in an aspect of trade within a liberalized framework is generally a pro-active measure, which brings to bear the willingness and ability of the governments to face trading issues squarely in the eye. However, any agreement for trading benefits would be ineffective without the element of competition, both between enterprises and between States. The essential requisite for success in trading relations is competition, which in turn leads to national prosperity. A free trade agreement is merely the catalyst in the process.

Success at competition in air transport is all about creating value over costs incurred. Here, value is what the customer pays for, whether in terms of service or quality of performance. With this strategy of V > C (where V is value and C is cost) the airline achieves a global competitive strategy which in turn gives the carrier the ability to differentiate its product and innovate, building a global network. One of the main ingredients to this competitive economic recipe is the recruitment of skilled executives who are committed to their job and bring with them prior experience of substance and value. They could leverage the existing market to meet demand and compete with others in the business. An example is IMAX, which in 1990 showed documentaries on their large screen but leveraged the existing cinema market by introducing blockbuster movies in 1995, which proved to be a huge disruptive innovation and success. Finally, with this approach, air carriers can prioritize foreign markets they wish to serve, select the best approach to serve their customers and adapt their products to customer needs.

Appendix A: Relevant Provisions of Treaty on the Functioning of the European Union on State Aid

Aids Granted by States

Article 107

(ex Article 87 TEC)

1. Save as otherwise provided in the Treaties, any aid granted by a Member State or through State resources in any form whatsoever which distorts or threatens to distort competition by favouring certain undertakings or the production of certain goods shall, in so far as it affects trade between Member States, be incompatible with the internal market.
2. The following shall be compatible with the internal market:
   (a) aid having a social character, granted to individual consumers, provided that such aid is granted without discrimination related to the origin of the products concerned;
   (b) aid to make good the damage caused by natural disasters or exceptional occurrences;
   (c) aid granted to the economy of certain areas of the Federal Republic of Germany affected by the division of Germany, in so far as such aid is required in order to compensate for the economic disadvantages caused by that division. Five years after the entry into force of the Treaty of Lisbon, the Council, acting on a proposal from the Commission, may adopt a decision repealing this point.
3. The following may be considered to be compatible with the internal market:
   (a) aid to promote the economic development of areas where the standard of living is abnormally low or where there is serious underemployment, and of the regions referred to in Article 349, in view of their structural, economic and social situation;

R. Abeyratne, *Competition and Investment in Air Transport*,
DOI 10.1007/978-3-319-24372-6

(b) aid to promote the execution of an important project of common European interest or to remedy a serious disturbance in the economy of a Member State;
(c) aid to facilitate the development of certain economic activities or of certain economic areas, where such aid does not adversely affect trading conditions to an extent contrary to the common interest;
(d) aid to promote culture and heritage conservation where such aid does not affect trading conditions and competition in the Union to an extent that is contrary to the common interest;
(e) such other categories of aid as may be specified by decision of the Council on a proposal from the Commission.

Article 108

(ex Article 88 TEC)

1. The Commission shall, in cooperation with Member States, keep under constant review all systems of aid existing in those States. It shall propose to the latter any appropriate measures required by the progressive development or by the functioning of the internal market.
2. If, after giving notice to the parties concerned to submit their comments, the Commission finds that aid granted by a State or through State resources is not compatible with the internal market having regard to Article 107, or that such aid is being misused, it shall decide that the State concerned shall abolish or alter such aid within a period of time to be determined by the Commission.

If the State concerned does not comply with this decision within the prescribed time, the Commission or any other interested State may, in derogation from the provisions of Articles 258 and 259, refer the matter to the Court of Justice of the European Union direct.

On application by a Member State, the Council may, acting unanimously, decide that aid which that State is granting or intends to grant shall be considered to be compatible with the internal market, in derogation from the provisions of Article 107 or from the regulations provided for in Article 109, if such a decision is justified by exceptional circumstances. If, as regards the aid in question, the Commission has already initiated the procedure provided for in the first subparagraph of this paragraph, the fact that the State concerned has made its application to the Council shall have the effect of suspending that procedure until the Council has made its attitude known.

If, however, the Council has not made its attitude known within 3 months of the said application being made, the Commission shall give its decision on the case.

3. The Commission shall be informed, in sufficient time to enable it to submit its comments, of any plans to grant or alter aid. If it considers that any such plan is

not compatible with the internal market having regard to Article 107, it shall without delay initiate the procedure provided for in paragraph 2. The Member State concerned shall not put its proposed measures into effect until this procedure has resulted in a final decision.

4. The Commission may adopt regulations relating to the categories of State aid that the Council has, pursuant to Article 109, determined may be exempted from the procedure provided for by paragraph 3 of this Article.

Article 109

(ex Article 89 TEC)
The Council, on a proposal from the Commission and after consulting the European Parliament, may make any appropriate regulations for the application of Articles 107 and 108 and may in particular determine the conditions in which Article 108 (3) shall apply and the categories of aid exempted from this procedure.

Appendix B: Relevant Competition Provisions of the Clayton Act

§ 2 Clayton Act, 15 U.S.C. §§ 13(2)

Discrimination in price, services, or facilities

(a) Price; selection of customers
It shall be unlawful for any person engaged in commerce, in the course of such commerce, either directly or indirectly, to discriminate in price between different purchasers of commodities of like grade and quality, where either or any of the purchases involved in such discrimination are in commerce, where such commodities are sold for use, consumption, or resale within the United States or any Territory thereof or the District of Columbia or any insular possession or other place under the jurisdiction of the United States, and where the effect of such discrimination may be substantially to lessen competition or tend to create a monopoly in any line of commerce, or to injure, destroy, or prevent competition with any person who either grants or knowingly receives the benefit of such discrimination, or with customers of either of them: Provided That nothing herein contained shall prevent differentials which make only due allowance for differences in the cost of manufacture, sale, or delivery resulting from the differing methods or quantities in which such commodities are to such purchasers sold or delivered: Provided, however, That the Federal Trade Commission may, after due investigation and hearing to all interested parties, fix and establish quantity limits, and revise the same as it finds necessary, as to particular commodities or classes of commodities, where it finds that available purchasers in greater quantities are so few as to render differentials on account thereof unjustly discriminatory or promotive of monopoly in any line of commerce; and the foregoing shall then not be construed to permit differentials based on differences in quantities greater than those so fixed and established: And provided further, That nothing herein contained shall prevent persons engaged in selling goods, wares, or merchandise in commerce from selecting their own customers in bona fide transactions and not in restraint of trade: And provided further, That nothing

R. Abeyratne, *Competition and Investment in Air Transport*,
DOI 10.1007/978-3-319-24372-6

herein contained shall prevent price changes from time to time where in response to changing conditions affecting the market for or the marketability of the goods concerned, such as but not limited to actual or imminent deterioration of perishable goods, obsolescence of seasonal goods, distress sales under court process, or sales in good faith in discontinuance of business in the goods concerned.

(b) Burden of rebutting prima-facie case of discrimination
Upon proof being made, at any hearing on a complaint under this section, that there has been discrimination in price or services or facilities furnished, the burden of rebutting the prima-facie case thus made by showing justification shall be upon the person charged with a violation of this section, and unless justification shall be affirmatively shown, the Commission is authorized to issue an order terminating the discrimination: Provided, however, That nothing herein contained shall prevent a seller rebutting the prima-facie case thus made by showing that his lower price or the furnishing of services or facilities to any purchaser or purchasers was made in good faith to meet an equally low price of a competitor, or the services or facilities furnished by a competitor.

(c) Payment or acceptance of commission, brokerage, or other compensation
It shall be unlawful for any person engaged in commerce, in the course of such commerce, to pay or grant, or to receive or accept, anything of value as a commission, brokerage, or other compensation, or any allowance or discount in lieu thereof, except for services rendered in connection with the sale or purchase of goods, wares, or merchandise, either to the other party to such transaction or to an agent, representative, or other intermediary therein where such intermediary is acting in fact for or in behalf, or is subject to the direct or indirect control, of any party to such transaction other than the person by whom such compensation is so granted or paid.

(d) Payment for services or facilities for processing or sale
It shall be unlawful for any person engaged in commerce to pay or contract for the payment of anything of value to or for the benefit of a customer of such person in the course of such commerce as compensation or in consideration for any services or facilities furnished by or through such customer in connection with the processing, handling, sale, or offering for sale of any products or commodities manufactured, sold, or quality, and quantity; to sell, or contract to sell, goods in any part of the United States at prices lower than those exacted by said person elsewhere in the United States for the purpose of destroying competition, or eliminating a competitor in such part of the United States; or, to sell, or contract to sell, goods at unreasonably low prices for the purpose of destroying competition or eliminating a competitor.

Any person violating any of the provisions of this section shall, upon conviction thereof, be fined not more than $5000 or imprisoned not more than 1 year, or both.

Cooperative association; return of net earnings or surplus, 15 U.S.C. § 13b

Nothing in sections 13 to 13b and 21a of this title shall prevent a cooperative association from returning to its members, producers, or consumers the whole, or any part of, the net earnings or surplus resulting from its trading operations, in proportion to their purchases or sales from, to, or through the association.

Exemption of non-profit institutions from price discrimination provisions, 15 U.S.C. § 13c

Nothing in sections 13 to 13b and 21a of this title, shall apply to purchases of their supplies for their own use by schools, colleges, universities, public libraries, churches, hospitals, and charitable institutions not operated for profit.

§ 3 Clayton Act, 15 U.S.C. § 14

Sale, etc., on agreement not to use goods of competitor It shall be unlawful for any person engaged in commerce, in the course of such commerce, to lease or make a sale or contract for sale of goods, wares, merchandise, machinery, supplies, or other commodities, whether patented or unpatented, for use, consumption, or resale within the United States or any Territory thereof or the District of Columbia or any insular possession or other place under the jurisdiction of the United States, or fix a price charged therefor, or discount from, or rebate upon, such price, on the condition, agreement, or understanding that the lessee or purchaser thereof shall not use or deal in the goods, wares, merchandise, machinery, supplies, or other commodities of a competitor or competitors of the lessor or seller, where the effect of such lease, sale, or contract for sale or such condition, agreement, or understanding may be to substantially lessen competition or tend to create a monopoly in any line of commerce.

Appendix C: The Sherman Antitrust Act (1890)

Section 1. Trusts, etc., in restraint of trade illegal; penalty
Every contract, combination in the form of trust or otherwise, or conspiracy, in restraint of trade or commerce among the several States, or with foreign nations, is declared to be illegal. Every person who shall make any contract or engage in any combination or conspiracy hereby declared to be illegal shall be deemed guilty of a felony, and, on conviction thereof, shall be punished by fine not exceeding $10,000,000 if a corporation, or, if any other person, $350,000, or by imprisonment not exceeding 3 years, or by both said punishments, in the discretion of the court.

Section 2. Monopolizing trade a felony; penalty
Every person who shall monopolize, or attempt to monopolize, or combine or conspire with any other person or persons, to monopolize any part of the trade or commerce among the several States, or with foreign nations, shall be deemed guilty of a felony, and, on conviction thereof, shall be punished by fine not exceeding $10,000,000 if a corporation, or, if any other person, $350,000, or by imprisonment not exceeding 3 years, or by both said punishments, in the discretion of the court.

Section 3. Trusts in Territories or District of Columbia illegal; combination a felony
Every contract, combination in form of trust or otherwise, or conspiracy, in restraint of trade or commerce in any Territory of the United States or of the District of Columbia, or in restraint of trade or commerce between any such Territory and another, or between any such Territory or Territories and any State or States or the District of Columbia, or with foreign nations, or between the District of Columbia and any State or States or foreign nations, is declared illegal. Every person who shall make any such contract or engage in any such combination or conspiracy, shall be deemed guilty of a felony, and, on conviction thereof, shall be punished by fine not exceeding $10,000,000 if a corporation, or, if any other person, $350,000, or by imprisonment not exceeding 3 years, or both said punishments, in the discretion of the court.

R. Abeyratne, *Competition and Investment in Air Transport*,
DOI 10.1007/978-3-319-24372-6

Section 4. Jurisdiction of courts; duty of United States attorneys; procedure
The several district courts of the United States are invested with jurisdiction to prevent and restrain violations of sections 1–7 of this title; and it shall be the duty of the several United States attorneys, in their respective districts, under the direction of the Attorney General, to institute proceedings in equity to prevent and restrain such violations. Such proceedings may be by way of petition setting forth the case and praying that such violation shall be enjoined or otherwise prohibited. When the parties complained of shall have been duly notified of such petition the court shall proceed, as soon as may be, to the hearing and determination of the case; and pending such petition and before final decree, the court may at any time make such temporary restraining order or prohibition as shall be deemed just in the premises.

Section 5. Bringing in additional parties
Whenever it shall appear to the court before which any proceeding under section 4 of this title may be pending, that the ends of justice require that other parties should be brought before the court, the court may cause them to be summoned, whether they reside in the district in which the court is held or not; and subpoenas to that end may be served in any district by the marshal thereof.

Section 6. Forfeiture of property in transit
Any property owned under any contract or by any combination, or pursuant to any conspiracy (and being the subject thereof) mentioned in section 1 of this title, and being in the course of transportation from one State to another, or to a foreign country, shall be forfeited to the United States, and may be seized and condemned by like proceedings as those provided by law for the forfeiture, seizure, and condemnation of property imported into the United States contrary to law.

Section 6a. Conduct involving trade or commerce with foreign nations
Sections 1–7 of this title shall not apply to conduct involving trade or commerce (other than import trade or import commerce) with foreign nations unless:

(1) such conduct has a direct, substantial, and reasonably foreseeable effect—
(A) on trade or commerce which is not trade or commerce with foreign nations, or on import trade or import commerce with foreign nations; or
(B) on export trade or export commerce with foreign nations, of a person engaged in such trade or commerce in the United States; and
(2) such effect gives rise to a claim under the provisions of sections 1–7 of this title, other than this section. If sections 1–7 of this title apply to such conduct only because of the operation of paragraph (1)(B), then sections 1–7 of this title shall apply to such conduct only for injury to export business in the United States.

Section 7. "Person" or "persons" defined
The word "person", or "persons", wherever used in sections 1–7 of this title shall be deemed to include corporations and associations existing under or authorized by the laws of either the United States, the laws of any of the Territories, the laws of any State, or the laws of any foreign country.

Index

A

ACI. *See* Airports Council International (ACI)
Aeronautical law, 2
Aerospace travel
  arms race, outer space, 18
  asteroids, 19
  commercial aerospace transport market, 15
  delimitation and demilitarization, 16
  economic theory, 19
  functionalist and spatialist approach, 16
  general and complete disarmament, 17
  international cooperation, outer space
    disaster management and emergency, 25
    education and training, 22
    friendly relations, 21
    geostationary orbit, 23
    global navigation satellite systems, 25
    legal problems, 20
    member States, 21
    mutual cooperation and assistance, 22
    peaceful purposes, 24
    remote earth resources, 21
    satellite-borne television, 21
    space objects, collisions of, 23
    transparency and confidence-building measures, 23
    UNCOPUOS, 20
  outer space, definition and the delimitation of
    air law and space law, 26
    Antarctic Treaty, 29–31
    commercial activities/services, 29
    demarcation, 26
    functionalist approach, 26
    maritime analogy, 31–35
    national sovereignty, 28
    scientific research, 28
    spatial approach, 27
  protectionism, 15
  space law, 16
  suborbital flights and space tourism, 18
  weaponization, outer space, 17
Aggressive unilateralism, 39
Airbus Industrie, 98
Airline Pilots Association (ALPA), 207
Airports Council International (ACI), 60, 219
Air Transportation Competition Act (Deregulation Act), 118
Air Transportation Safety and System Stabilization Act, 119
Air Transport Committee, 42, 43
Air transport infrastructure
  in China, 210
  FAA, 211
  Global Infrastructure Index, 210
  in Indonesia, 210
  Oxford Economics, 210
Air transport law
  aerial military activity, 3
  aeronautical law, 2
  alternative fuel technology, 6
  challenges
    commercial space transport, 9–10
    ICAO, role of, 9
    market access, 8–9
    nationalism and sovereignty, 6–7
    passenger rights, 11–13
  civil aviation, 3
  economic power, 3
  economic proactivity, 5
  international law and politics, 2

R. Abeyratne, *Competition and Investment in Air Transport*,
DOI 10.1007/978-3-319-24372-6

Air transport law (*cont.*)
- liberalization of, 1
- military warfare, 2, 3
- protectionism, 4
- public international law, 3
- self-induced stagnation, 5
- space law, 2, 3
- WEC, 5

ALPA. *See* Airline Pilots Association (ALPA)
Antarctic Treaty, 29–31
ASAM. *See* ASEAN Single Aviation Market (ASAM)
ASEAN Air Transport Integration Project (AATIP), 153
ASEAN Economic Community (AEC), 151
ASEAN Single Aviation Market (ASAM)
- AEC scheme and AEC, 152
- air transport economics, in Indonesia
  - branding strategy, 161
  - and China, 163
  - domestic traffic, 159
  - Garuda Indonesian and Merpati Nusantara airlines, 158
  - Law No 1 of 2009, 162
  - Law No. 1 of 2009 on Aviation (The Aviation Law), 159
  - technical requirements, 160
- ASEAN airlines, 155
- goal, 151

Attribution
- Airline Information and Thinkink, 212
- Article 8, of ILC principles, 212
- Classical theory, 214
- International Court of Justice, 213
- Maffezini case, 213
- Sixth Air Transport Conference of ICAO, 213
- State's investments, economic growth, 212

B

Boeing, 99, 148
Brussels South Charleroi Airport (BSCA), 90

C

CAEP. *See* Committee on Aviation Environment Protection (CAEP)
CANSO. *See* Civil Air Navigation Services Organization (CANSO)
CBDR. *See* Common but Differentiated Responsibilities (CBDR)
Chicago Convention, 6–12, 43, 50, 55, 64, 66, 74, 75, 116, 123, 135, 209
Civil Air Navigation Services Organization (CANSO), 59–60
Clayton Act
- 15 U.S.C. §§ 13(2), 227–229
- 15 U.S.C. § 14, 229

Codes of conduct (COCs), 12
Committee on Aviation Environment Protection (CAEP), 40
Committee on Foreign Investment in the United States (CFIUS), 171
Committee on the Peaceful Uses of Outer Space of the United Nations (COPUOS), 15, 17, 20, 22, 23, 28, 29, 33
Common but Differentiated Responsibilities (CBDR), 38, 43, 46, 49
Competition
- aerospace travel (*see* Aerospace travel)
- air carriers, 77
- anti competitive practices, 76
- commercial aviation, 74
- company's performance, 73
- complacency, 78
- complementary services, 74
- deregulated system, 74
- environmental protection, 78
- export subsidies, 87
- free trade, 73
- ground handling and precise engineering, 75
- ICC, 74
- and innovation
  - customer-base, 107
  - disruptive innovation, 108
  - nascent low cost carrier model, 107
  - Skybus, 108
  - wing-warping method, 109
- international legislative structure, 76
- issues at stake
  - actionable subsidies, 94
  - Airbus dispute, 92
  - complaints and counter-complaints, 97
  - DSB, 96
  - GATT/1947 system, 93
  - international order, 93
  - Marrakesh Declaration, 95
  - Tokyo Round Subsidies Code, 92
  - WTO dispute settlement system, 94
- ITO, 88
- liberalization, 74
- market access, 73
- market economics and air transport
  - contagion effect, 104

disruptive innovation, 106
employment growth, 106
free market, 105
IMF, 103
political and economic restrictions, 105
World Bank, 103
mathematical theories, 102–103
past, present and future, 98–99
principles of
air traffic rights, 87
antitrust laws, 82
aviation, 79
block exemptions, 81
brand and customer expectation, 79
Clayton Act, 82
commercial aviation and market access, 85
Common Market, 81
competitive strategy, 78
deregulated global airline industry, 80
GATS, 83
ICAO, 83
inflation, 79
international agreements, 84
international air transport, 83
leisure market, 80
multilateralism, 84
open skies policy, 86
preferential measures, 86
services negotiations, 84, 85
Sherman Act, 82
stringent competitor analysis, 79
signatories, 89
strategic airline management, 77
trade practices and state aid
Australian Trade Practices Act, 91
Clayton Act, 91
Common Market, 89
Competition Act, 90
Ryanair Case, 90
Treaty of Rome, 90
trans-continental flight, 75
UAE carriers, 99–101
WTO dispute settlement mechanism, 88
Competition Laws
Article 86, 199
Article 102, 200
Article 17, Anti-Monopoly Law of China, 194
Article 82, The Treaty Establishing the European Community, 199
ATConf/5, 205
ATConf/6, 205
in the Czech Republic, 195
in Estonia, 195
European Court of Justice, 201
in Germany, 195–196
ICAO model, 204
Indian Competition Act, 194
market liberalization, 204
Model Law, 194
Regulation No 17/62, 200
Sixth Air Transport Conference of ICAO, 203
in South Africa, 194
state aid, 196–199
in United States, 201–202
in Zambia, 195
Competitive advantage
aviation, 132
connectivity, air transport
geographic locations, 136
liberalization, 134
RPKs, 134
tourism, 135
corporate culture, 133
corporate strategy, 132
employee empowerment, 133
free market, 131
globalization, 131
hubbing, 136–138
innovation
disruptive innovation, 139, 140
efficiency innovation, 139
fair competition, 141
The Game Theory, 141
market creating innovation, 139
non-consumer market, 139
perfect competition, 142
sustaining innovation, 139
unfair competition, 142
workable competition, 142

D
Department of Transportation Office of Inspector General (DOT-OIG), 65
Dispute Settlement Body (DSB), 96, 97
Disruptive innovation, 102, 103, 106–109, 133, 139–141, 191, 222
DSB. *See* Dispute Settlement Body (DSB)
Dubai International Airport, 208

E
EAG. *See* Environment Advisory Group (EAG)

EEZ. *See* Exclusive economic zone (EEZ)
Emirates Airlines, 192, 210
Emissions Trading Scheme (ETS), 38, 39, 41
Emission unit criteria (EUC), 44
Environment Advisory Group (EAG), 40–45
Environment, competition and
  aircraft emissions, 37
  CBDR, 38
  embarkation taxes, 41
  emerging economies, 37
  ETS, 39, 41
  falsiability, 40
  ICAO
    CAEP process, 44–45
    EAG, 42–43
  international aviation community, 39
  Rio Declaration, 38
  SCRC, 38
  Strawman approach, 39, 41, 42, 45–47
ETS. *See* Emissions Trading Scheme (ETS)
Exclusive economic zone (EEZ), 32

F
Federal Aviation Administration (FAA), 65, 211
Filing of Differences Task Force (FDTF), 68
Foreign Trade Antitrust Improvements Act, 118

G
Game Theory, 102
GATT. *See* General Agreement on Tariffs and Trade (GATT)
General Agreement of Trade in Services (GATS), 83, 84, 121, 122, 124, 125, 128
General Agreement on Tariffs and Trade (GATT), 84, 85, 87, 88, 93, 95–97, 118, 125, 165–168, 175, 178, 184, 185
GesetzgegenWettbewerbsbeschränkungen (GWB), 90
Global Aviation Dialogues (GLADs), 44

I
IAEA. *See* International Atomic Energy Agency (IAEA)
IATA. *See* International air Transport Association (IATA)
ICAO. *See* International Civil Aviation Organization (ICAO)
ICC. *See* International Chamber of Commerce (ICC)
IFALPA. *See* International federation of Airline Pilots Associations (IFALPA)
Impacts and Science Group (ISG), 45
Indian Sub Continent (ISC), 140
International Air Transport Association (IATA), 1, 6, 8, 75, 101, 129, 134, 155, 160, 192, 203, 218–220
International Atomic Energy Agency (IAEA), 22
International Chamber of Commerce (ICC), 74, 116, 157, 221
International Civil Aviation Organization (ICAO)
  aircraft emissions, 38
  air transport, 9
  Air Transport Regulation Panel, 217
  ATConf/6, 2013, 216
  Chicago Convention, 217
  consumer protection, 12
  ICAO Council, 217–218
  perspectives, 121–123
  Strawman considerations, 48–51
  work of, 42–45
International federation of Airline Pilots Associations (IFALPA), 64
International Monetary Fund (IMF), 103
International Telecommunications Union (ITU), 21, 23
International Trade Organization (ITO), 88
ISG. *See* Impacts and Science Group (ISG)
ITU. *See* International Telecommunications Union (ITU)

J
Just Culture, 60

L
Long-range Trans-boundary Air pollution (LRTAP), 50

M
Maritime analogy
  Antarctic Treaty, 31
  delimitation, 34
  economic exploitation and exploration, 33
  EEZ, 32, 33
  enforcement model, 35

International Seabed Authority, 32
Moon Treaty, 31
Outer Space Treaty, 31
space transportation systems, 34
territorial sea, 32
Market access
debate
connectivity, 116
Etihad claims, 114
global regulatory framework, 117
ICAO, 117
international economic law, 116
liberalization, 115
QANTAS, 115
U.S. Open Skies policy, 114
Europe, 119–121
subsidies and WTO
funding mechanism, 129
globalization and information revolution, 129
international air transportation, 128
national policy, 126
open skies agreements, 127
SCM Agreement, 126
traffic light approach, 128
WTO Panel, 127
United States (US), 118–119
WTO
ICAO perspectives, 121–123
misconceptions, 124–125
Market Based Measures (MBM) Scheme
aviation emissions, 40
GLADs, 44
ICAO, 42
Strawman approach, 47–48
Monitoring, reporting and verification (MRV), 44
Montreal Convention, 62
Most Favoured Nation (MFN), 125
MRV. *See* Monitoring, reporting and verification (MRV)
Multilateral Agreement on Air Services (MAAS), 152
Multilateral Agreement on the Full Liberalization of Passenger Services (MAFLPAS), 152

N

NAFTA Free Trade Agreement (NAFTA)
broader issue of subsidies, 167
Canada
Article III:4 of GATT, 168
foreign investment, 167–168
Investment Canada Act, 169
trading partners, 168
dispute settlement, 167
GATT, Article XXIV, 166
historical and economic aspects
Article XXIV of GATT, 178
Canada's decision, 178
Canadian market, 177
Framework Trade Agreement, 179
free trade, 179
Free Trade Agreement, 178
Mexico's inclusion, 179
Mexico-United States relationship, 180
tariff and non tariff barriers, 176
laws and policies, 167
legal status, 181–184
Mexico
exchange control laws, 175
foreign currency income, 176
foreign investment law, 173, 174
foreign investors, 173
free market, 176
laws and regulations, 172
parties rights and obligations
Article 1114 (1), 185
Article 1114(2), 185
Article 1406, 186
Article 1102 (1) and (2), 186
Article III of GATT, 185
investors and financial institutions, 186
protection and security, 187
Provisional International Court of Justice (PCIJ), 186
performance requirements, 167
principle of national treatment, 166
United States
CFIUS, 171
Clayton Act of 1981, 172
federal laws, 172
foreign investment, 170
open door policy, 171

O

O&D approach, 192
OECD. *See* Organization for Economic Cooperation and Development (OECD)
Open skies agreement, 118, 202
anti-trust exigencies, 147
ASEAN and Indonesia

Open skies agreement (*cont.*)
ASEAN Economic Community (ACE), 151
Garuda, 151
Indonesia's archipelagic topography, 150
liberalization policy, 152
MAAS, 152
MAFLPAS Protocol, 152
bilateral air services agreements, 149
Brattle Report, 146
commercial considerations, 147
convergence of regulatory principles, 148
EU, 145
EU-US relations, 147
global economy
AATIP, 153
AEC, 154
air transport, 156
ASAM and Air Transport Economics, 158–163
free trade, 156
ICC, 157
off-shore base, 157
Oxford Economics, 157
political and economic restriction, 158
tourism, 154
UNWTO, 154
TCAA concept, 148
Organization for Economic Cooperation and Development (OECD), 4, 53, 98, 132, 151, 191
Outer Space Treaty, 17–19, 26, 29–31
Oxford Economics, 100, 210, 212, 220

P
Partial Test Ban Treaty, 17
Probability Theory, 102

Q
QATAR Airways, 193

R
Revenue passenger kilometres (RPKs), 134

S
Safety, competition and
impact
aircraft operators, 54
audit follow-up missions, 55
economic benefits, 55
international civil aviation, 54
liberalization, 54
OECD, 53
safety information
Conference, considerations of, 56–57
ICAO, 55
Recommendations, 57–58
SARPs, 56
web-based system, 56
second high level safety Conference
consumer protection, 69–72
dissemination and sharing of, 66–68
GermanWings, 63–66
legal issues, 61–63
protection, safety information, 58–60
SARPs. *See* Standards and Recommended Practices (SARPs)
SCRC. *See* Special Circumstances and Respective Capabilities (SCRC)
SDRs. *See* Special Drawing Rights (SDRs)
Second high level safety Conference
consumer protection
guidelines, 70
ICAO Council, 72
provision of services, 71
dissemination and sharing of, 66–68
GermanWings
code of ethics, 64
FAA, 65
Japan Airlines, 63
Malaysian Flight MH 370, 63
medical certification, 64
Mozambique Airlines aircraft, 64
Prozac, 65
regulations, 66
Royal Air Maroc flight, 64
SilkAir Flight, 63
legal issues
Air Asia flight, 62
Contracting States, 69
Dutch Safety Board, 61
Global Aviation Safety Plan, 69
"ICAO" treaties, 61
performance-based regulations, 68
SDRs, 62
standardization and harmonization, 63
protection, safety information
CANSO *Just Culture,* 60
disciplinary and compliance purposes, 59

EUROCONTROL, 59
ICAO Secretariat, 58, 60
Sherman Antitrust Act, 118, 231–232
Silicon Valley, 219
Social Security Administration Office of Inspector General (SSA-OIG), 65
Space Activities Act, 34
Special Circumstances and Respective Capabilities (SCRC), 38, 43, 46, 49
Special Drawing Rights (SDRs), 62
Standards and Recommended Practices (SARPs), 56, 61, 68
State aid, EU
Article 107, 223–224
Article 108, 224–225
Article 109, 225
Strawman approach
falsification, 46
ICAO
Assembly Resolution A38-18, 48
climate change, 49
compliance mechanism, 50
law making process, 50
SCRC, 49
welfarist approach, 50
MBM scheme, 47–48
Popper's theory, 46
tentative draft proposal, 45
Subsidies and Countervailing Measures (SCM) Agreement, 8, 94, 126–129
Sukarno Hatta Airport, 152

T
TAP Air Portugal tax, 119
Template Air Services Agreements (TASAs), 123

U
United Nations Convention on the Law of the Sea (UNCLOS), 31–33
United Nations Educational, Scientific and Cultural Orga nization (UNESCO), 22
United Nations Framework Convention on Climate Change (UNFCCC), 40
Universal Safety Oversight Audit Programme (USOAP), 69

V
Vienna Convention on the Law of Treaties, 182

W
Warsaw Convention, 62
WEC. *See* World Energy Council (WEC)
WMO. *See* World Meteorological Organization (WMO)
World Energy Council (WEC), 5
World Meteorological Organization (WMO), 22
World Trade Organization (WTO), 8, 117

MIX
Papier aus verantwortungsvollen Quellen
Paper from responsible sources
FSC® C105338

If you have any concerns about our products,
you can contact us on
ProductSafety@springernature.com

In case Publisher is established outside the EU,
the EU authorized representative is:
Springer Nature Customer Service Center GmbH
Europaplatz 3, 69115 Heidelberg, Germany

Printed by Libri Plureos GmbH
in Hamburg, Germany